A theoretical physicist looks at

THE SUN

JRBreton

A theoretical physicist looks at
THE SUN
by JRBreton

Published by:
The Foundation for Theoretical Physics
3 Apple Tree Lane
Walpole, MA 02081–2301

Web address: FoundationForTheoreticalPhysics.org
email address: theoretical.physics.books@gmail.com

Copies of this book and other offerings of the Foundation may be obtained online from the Foundation's website and also from amazon.com and similar sites.

ISBN, print ed. **978-1-63625-639-9**

First printing 2023

Printed in the United States of America

Library of Congress Control Number 2023917605

Table of Contents

Introduction.

Like anyone else, a theoretical physicist looking directly at the sun risks having his eyes burnt out. So that is not the way a theoretical physicist looks at the sun. How then?

Instead, armed with the many ideas and theorems of his science, he looks at the sun *intellectually*. And how does he do that?

To answer that question one has to investigate the many ideas and theorems of Theoretical Physics. These may be found in the foundational works.

Foundational literature

Theoretical Physics: The First Problem
ISBN 978-0-9844299–1-2
afterwards referenced as *The First Problem*
Theoretical Physics: The Second Problem
ISBN 978-0-9844299-3-6
afterwards referenced as *The Second Problem*
Theoretical Physics: The Third Problem
ISBN 978-0-9844299-5-0
afterwards referenced as *The Third Problem*

The first problem is elaborated in

tp1.1: An inquiry into the Foundations of the Science of Physics
ISBN 978-0-9844299-7-4
afterwards referenced as *tp1.1*
tp1.2: A continuing inquiry into the Foundations of the Science of Physics: Vector Algebra
ISBN 978-0-9844299-8-1
afterwards referenced as *tp1.2*

tp1.3: A continuing inquiry into the Foundations of the Science of Physics: Vector Calculus I
ISBN 978-0-9844299-9-8
afterwards referenced as *tp1.3*
tp1.4: A continuing inquiry into the Foundations of the Science of Physics: Vector Calculus II
ISBN 978-1-63625-652-8
afterwards referenced as *tp1.4*
tp1.5: A continuing inquiry into the Foundations of the Science of Physics: The Bridge Theorem
ISBN 978-1-63625-653-5
afterwards referenced as *tp1.5*

Additional information can be found in the following:

Step Functions and Product Rules
ISBN 978-0-9844299-2-9
Differentiation and Integration of Compound Functions
ISBN 978-0-9844299-4-3
Playing with Einstein
ISBN 978-0-9844299-0-5
Playing with the Big Bang
ISBN 978-0-9844299-6-7

Before turning our attention to the sun, we will do well to consider some preliminary topics the better to appreciate the language of Theoretical Physics.

Elementary epistemology

First a basic question: How do we know anything at all? Obviously we have knowledge, knowledge of many kinds. For our purposes, let us distinguish two different kinds: sensory and intellectual knowledge. They are distinguished as in the following table.

Sensory Knowledge	Intellectual knowledge.
effect of of sensory faculties	effect of an intellectual faculty
results in sensory truth	results in intellectual truth
has physical reality as its object	has both physical and ideal reality as its object
Immediate	Compound
Infallible[1]	Rational, and so fallible
awesome	awesome

Differences between Sensory and Intellectual Knowledge

We observe sensibly, we understand with ideas. We humans share with other animals the awesome power of the senses to provide sensory knowledge. The awesome power of the human intellect to provide ideas resides peculiarly in human beings. Language provides a bridge between the two powers.

Language is composed of sensible objects which are accepted not as such, but as pointers to ideas.

Ordinarily intellectual knowledge arises from sensory knowledge, but not always. Ideas may be conceived with no direct link to a sensory object.

Intellectual knowledge may be further distinguished according to the following table.

1. What we sense, we sense as a sensory impression. We may observe something, white, furry, moving, etc. We understand the object to be a cat, only after we have conceived the idea of a cat. The word "cat" is a sensible object which refers to the idea.

Technology	**Mathematical Science**	**Science of Reality**
utilitarian	not necessarily useful	not necessarily useful
develops new methods to meet new needs	permanent methods using non–contradictory axioms	permanent methods using axioms corresponding to reality
tolerates approximations	does not tolerate approximations	does not tolerate approximations
relies on measurements	does not rely on measurements	does not rely on measurements

Compare Accounting with Arithmetic. Although both disciplines use numbers, Accounting should be classed as a technology while Arithmetic should be classed as a Science. The arithmetical truth that there is no maximum positive integer offers no utility at all for the accountant.

Science & Technology has, in the popular mind, a somewhat sacrosanct meaning: namely, an activity which is both permanently truthful and useful. It is a good example of sloppy thinking by neglecting to make critical distinctions. Science & Technology is an oxymoron.

Much of what is considered Physics in our time is technology rather than science. Modern Physics relies on measurements and willingly accepts contradictions.
The attraction of a true science of Physics nevertheless prompts the proponents of Modern Physics to claim its mantle. Numerous attempts have been made to specify its axioms, none of which are satisfactory. Even the definition of Physics is not readily available.

Definition of the Science of Physics

For the theoretical physicist a definition of the science of Physics is given in *Theoretical Physics: The First Problem* as

Physics is:

the study of

reality observable as

extended objects

which may be

moving

or forcing.

That such a study can be made into a science strictly speaking is the burden of *The First Problem*.

The scientific study of the sun is a branch of the science of Physics which may be defined as

The Physics of the sun is:

the study of

the sun observable as

an extended object

which is

moving

and forcing.

The sun may be studied and appreciated in many other ways.

It figures in the religious literature of the ages.
It figures in the poetry of many languages.
It figures in medicine and health literature.
It figures in biology and as a necessity for earthly life.

None of these, valid as they are for human contemplation, is scientific, nor need they be.

Why study the sun? The technological answer is because of the utility of knowing its effects on our lives, or to evade deleterious effects it could have on us.

But why a scientific study? Because it is our nearest star and so best suited for such study.

The sun's nearest astral neighbor is Alpha-Centauri some 4.24 light-years distant. Not so far? Don't be deceived by a small number. It is 65,700 times more distant than the distance of the earth to the sun. To put the matter in earthly terms: If the earth were set on the 0 yard-line of a football field in New York City and the sun on the 92 yard-line, Alpha-Centauri would be placed in London England.

Transformation of Mathematics into Theoretical Physics

Mathematics, as such, is not suitable for physics. Why not?

For many reasons, one being the need for units, another references, another resolution.

As noted in tp1.1 mathematics is transformed into theoretical physics by first restricting a mathematical idea, and then expanding the result.

Units

Acknowledging the basic physical units of extension (L), motion (V), and force (F), Theoretical Physics addresses combinations like

$$L * L$$
$$V * V$$
$$F * F$$
$$L * V$$
$$L * F$$
$$V * F$$

Consequently ideas like area, energy, power are recognized by Theoretical Physics as ideas and not physical realities. Such ideas may be useful in describing and explaining observations, but must always be traceable to the elementary physical realities of extension, motion, and force.

References

To further the transformation of mathematical ideas into Theoretical Physics, a way is needed to identify change in the observed object.

First consider what is meant by a particle.

For Theoretical Physics a particle (**x**) is defined as follows.

Definition (particle)

Given

S(**x**) a set of objects with a topology

x a member of S(**x**)

VT(**x**) the set of all open sets containing **x**

then

A **particle** (**x**) is the intersection of all the subsets of VT(**x**)

end of definition

The definition of a topology can be found in Mathematics.

In Mathematics the intersection is a mathematical limit, whereas in Theoretical Physics the intersection is limited to the interest of the observer, or to the resolution of the observation.

The definition of a particle clearly refers to ideas, rather than objects. A particle is an *idea* of Theoretical Physics. It may only roughly correspond to an objective reality. The *resolution* desired by the physicist is key. An astronomer, for instance, may consider a star a particle, whereas a chemist might consider a tiny crystal a particle.

The set function, **x,** identifies extended physical matter in the universe by associating it uniquely with a set of locations in a three dimensional set of vectors designated as **V3**. With reference to a hypothetical universe, directly observable primary changes in matter can be described as functions of **x** symbolized as follows:

Observable	**At Rest**	**Afterwards**	**Idea**
extension	**x**	**r**(**x**)	location
motion	0	**v**(**x**)	velocity

Material/Local References

Consider the function, **r: V3**(**x**)⟶**V3**(**r**), denoted by **r**(**x**), which describes the change, if any, in the location of an observable particle, (**x**).

The function **r**(**x**) is called a set function because it connects two different kinds of sets: the location of matter in the universe to **V3.**

In Theoretical Physics two modes of observation are considered: local and material. In our first and coarse approach to solar physics only the local mode is considered. Later when more refined observations of the particles is investigated both modes will have be considered.

The first way of observation, **r**(**x**), is called the **material** reference. To distinguish the global transformation from a particular function the following symbology is used:

	Global	Local
Location	**r**(**x**) or **r**|**x**	**r**(**x1**) or **r**|**x1**

Observations

Physical observations are ordered, that is, one observation follows another. Any observation comes after or before another. Theoretical Physics proposes numbering the ordered observations as a real[1] variable, symbolized as *a*.

Usually concern is centered on a local perspective at observation *a1* with the particle **x1** as reference whose location at observation *a1* is **r1**(**x1**,a1).

---

1. For Theoretical Physics, quotient numbers are sufficient.

The following expresses a fundamental principle distinguishing both Physics and Theoretical Physics from Mathematics.

The Principle of Non–Collocation

For the same observation,
two different particles cannot occupy the same position.

Symbolically,

$$\{\mathbf{r}(\mathbf{x2},a1) = \mathbf{r}(\mathbf{x1},a1)\} \longrightarrow \mathbf{x2} \text{ is } \mathbf{x1}$$

Why does the principle of non-collocation matter?

As a first consequence of the principle of non–collocation, notice that in Theoretical Physics **r**(**x1**,a1) cannot be decomposed into

$$r1(\mathbf{x1},a1)*\mathbf{u1} + r2(\mathbf{x1},a1)*\mathbf{u2} + r3(\mathbf{x1},a1)*\mathbf{u3}$$

where the **ui** are an orthogonal set of directions, but rather

$$\mathbf{r}(\mathbf{x1},a1) = r1(\mathbf{y1},a1)*\mathbf{u1} + r2(\mathbf{y2},a1)*\mathbf{u2} + r3(\mathbf{y3},a1)*\mathbf{u3}$$

for some **y1**, **y2**, and **y3**, because **x1** at observation *a1* occupies location **r**(**x1**,a1), not r1∗**u1**.

We might reflect on thoughts that lie beyond and above Physics. There is another way to express the schema linking the physical world to the intellectual ideas. The physical universe is its matter. By His act of creation God materialized the idea of a three dimensional set of vectors. In so doing He gives these ideas a new material existence without in the least compromising them as ideas.

God's purpose is revealed by his Messiah, Jesus, who comes to spiritualize matter, that is, to give matter a new spiritual existence. Jesus does this by revealing Himself as God's incarnation, that is God becoming part of the material universe without in the least compromising His divinity. In this way creation returns to its source.

This knowledge can only come from God who freely reveals it. But once revealed we can marvel how well it comports with Theoretical Physics. The duality in Theoretical Physics is a reflection in the physical world of the centrality of Jesus Christ, who revealed Himself by his life, death, and resurrection as the unique divine/human duality. As such, He is the origin, the one who continues, and the destiny of the material universe. It is in Him that the "groanings"of material creation find resolution.

While the mystery of Jesus is valued especially for enlightenment of the human condition, it is not without light for the physical universe too.

Returning to our task: *tp1.2:* describes how vectorial plus and minus can be brought into the framework of Theoretical Physics with restrictions. Also by recognizing that vectors and be multiplied in three ways, and also divided, the same volume shows how vector multiplication and division can be brought into Theoretical Physics. As such Theoretical Physics also comprehends an algebra and consequently the ability to solve algebraic equations.

Syntax

The general syntax for operators in Theoretical Physics is

Operator rank [**reference**] **couple** (**f** | **condition; increment**).

For example The positive quadrant curl of a vector function is symbolized as **D3**[**r1**]∧(**f**(**r**);**dr**) where

Operator	is **D** for derivative
rank	is 3 for dimension
reference	is **r1** for location
couple	is ∧ for curl
f	is the function being differentiated
condition	is not expressed here in the symbol
increment	is **dr** a vector

For another example the invergence of a gradient of a function along a curve is symbolized as

I[r1,rm] • (**D[r,r+dr(r)]**∗(**f(r)** | CR;**dr**) | CR;**dr**)

where

Operator	is **I** for integral
rank	is inferred
reference	is [**r1**,**rm**] for a section along a curve
couple	is • for invergence
f	is (**D[r,r+dr(r)]**∗(**f(r)** | CR;**dr**)
	the gradient being integrated
condition	is CR the curve
increment	is **dr** a vector

As may be easily appreciated, a large number of combinations may be realized.

Coarse Considerations

Theoretical Physics accepts position and motion as observables, not something to infer by calculation.

We start with two particles and their positions **r**(**x1**,a) and **r**(**x2**,a), later to be identified as the earth and the sun.

The symbol **r**(**x1**,a) refers to a vector which designates the position **r** of the particle **x1** at some observation a. The vector has a direction and a magnitude with reference to some origin.

We are interested in **r**(**x2**,a) – **r**(**x1**,a). For **x2** the earth and **x1** the sun, the difference is the path of the earth around the sun measured as distances and directions
See the following diagram.

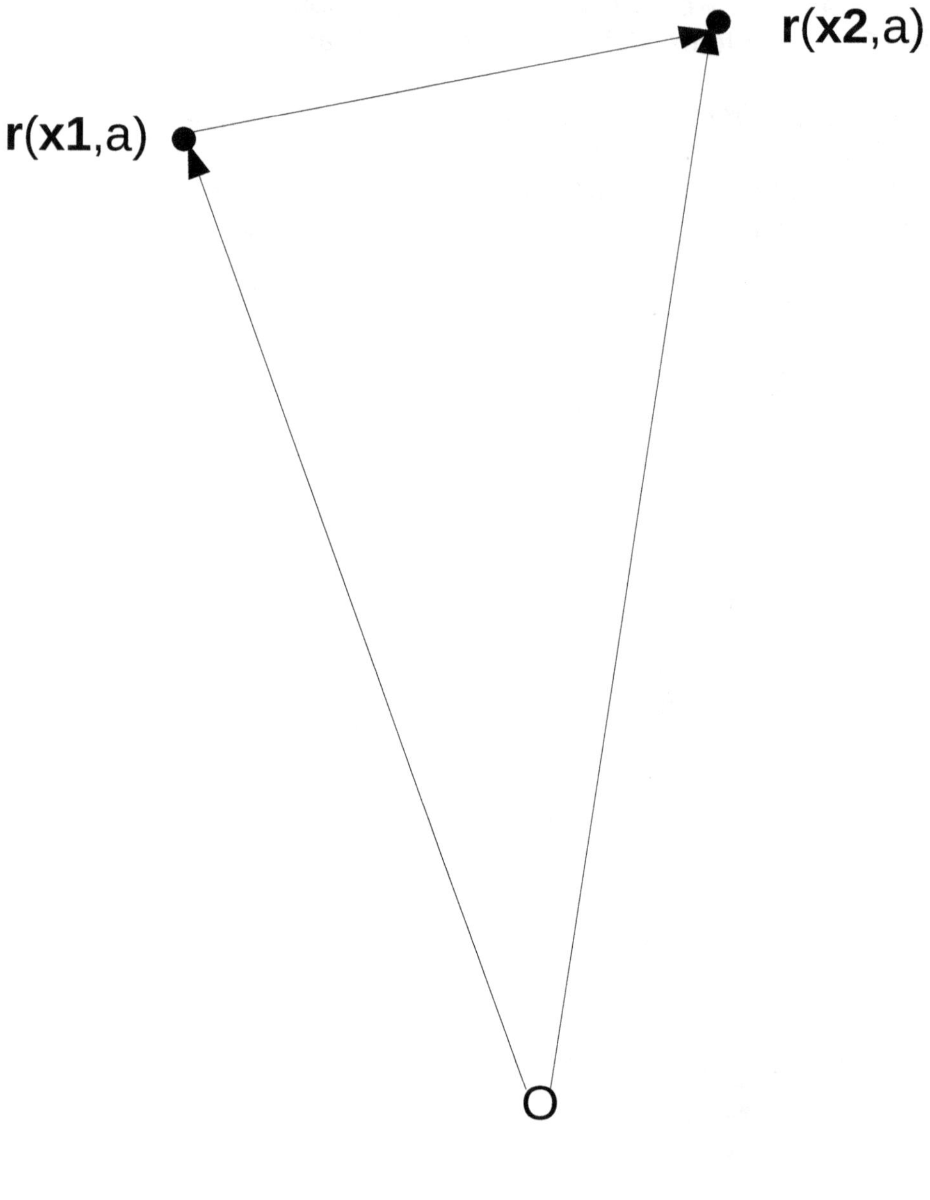

Next we observe the motions of the particles as well as their positions. Call the vector of the motion **cx**. The movement of a particle is defined as

$$\mathbf{cx} \equiv \lim (\mathbf{r}(\mathbf{x},a1+da) - \mathbf{r}(\mathbf{x},a1))/da \quad \text{as } da \rightarrow 0$$

Physically, both particles are moving, as in the diagram.

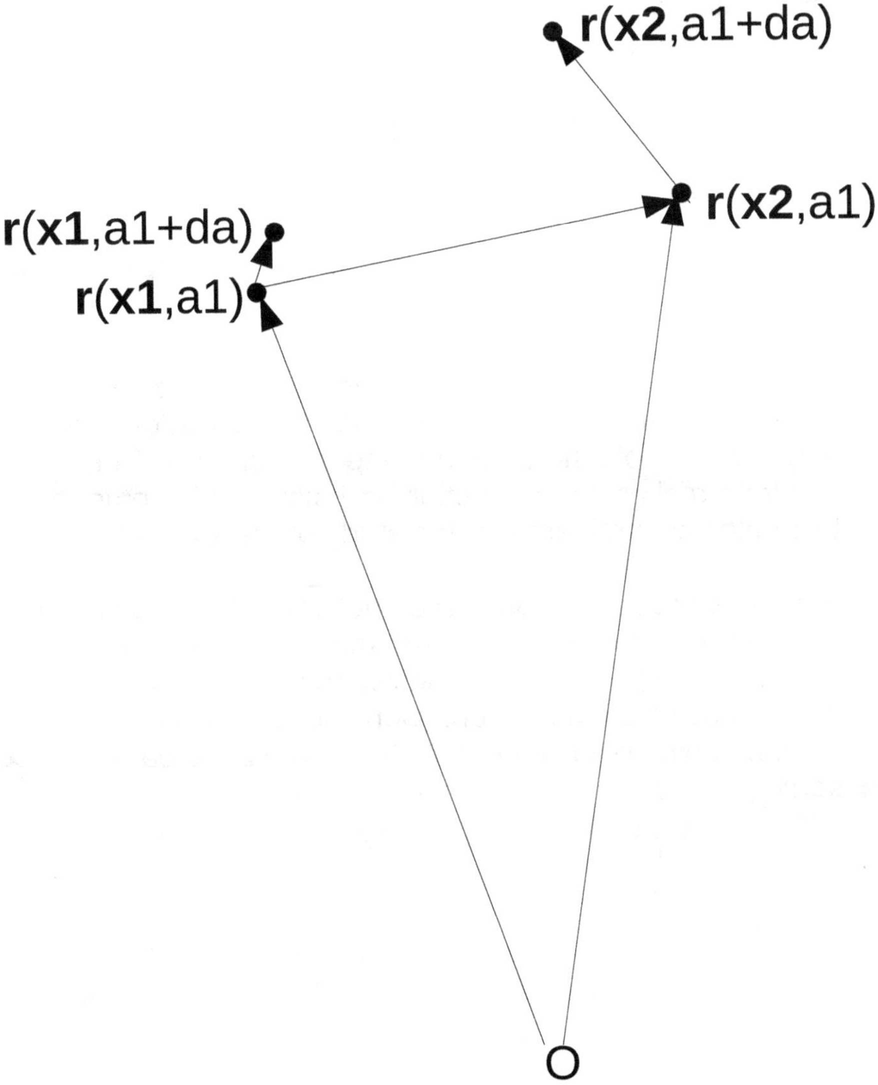

Both **x1** and **x2** may be moving at different speeds and in different directions. We are interested only in

cx(**x2,**a) – **cx**(**x1**,a).

(This condition equivalently makes **cx**(**x1**,a) = **0**.)

The relative positions of the two particles may be described as an integral function of observations

I[a0,a] (**cr**(**x1**,a);da)= **r**(**x1**,a) – **r**(**x1**,a0)

or as a function of time[1]

I[t0,t] (**vr**(**x1**,t);dt) = **r**(**x1**,t) – **r**(**x1**,t0)

Orbits

These integrals trace out the path of **x2** around **x1**, whether observed from **x1** or **x2**.

The actual measurement is a technological problem. Observation from the earth's surface has the additional problem of a rotating earth and the atmosphere. The accurate reckoning of direction and distance become major technological problems for the study of the sun.

From this outlook the difference between a science of the sun and a technology of the sun may be appreciated. The scientific viewpoint does not insist on a single orbit as the earth's path around the sun. Consider the following diagrams where **x1** represents the sun, **x2** the earth, and **x3** some other body like the planet Jupiter.

1. The relationship between observation and time is given in *The Second Problem*.

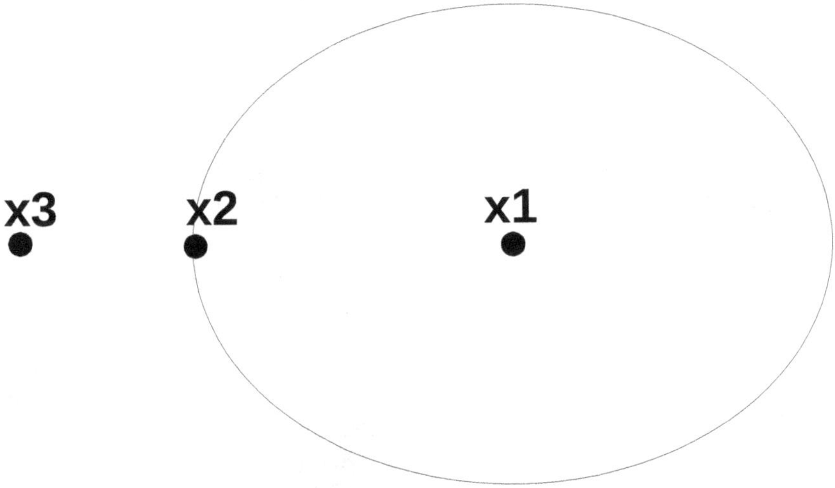

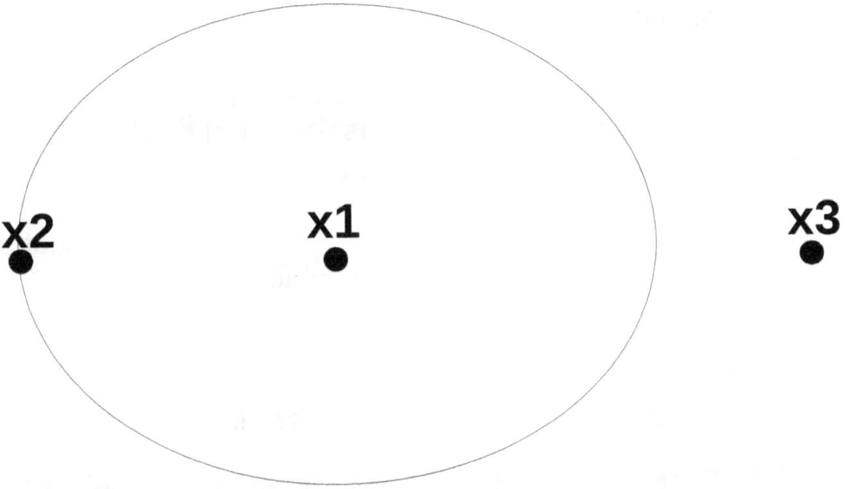

In the upper diagram the path of **x2** would be somewhat bowed away from **x1** toward **x3** because of gravity, whereas in the lower diagram the path would be bowed somewhat toward **x1**.

Solar Physics, as a science, observes the path of the earth around the sun as determined by all the influences on it, and concludes that the earth has not one but many different orbits around the sun, and these change from time to time.

Solar technology, on the other hand, has concentrated on calculating *the* orbit of the earth. A first effort was made by Nicolaus Copernicus (1473-1543) who assumed the orbit circular. He could thus posit two variables, the radius of the orbit and the velocity of the earth in the orbit. The Copernican model was found unable to match the observed data.

A second effort was made by Johannes Kepler (1571-1640) who assumed an elliptical orbit and thus was able to introduce a third variable into the model and allow for a varying velocity. This model was found useful for navigation and widely accepted. Isaac Newton (1643-1627) based his physics on the Keplerian model.

Technology produces something useful, else it is discarded. Insofar as it accepts contradictions, it cannot produce understanding.

True science, on the contrary, will not accept contradictions, and so can produce understanding which lasts.

Task

Distinguish solar science from solar technology.

Perihelion of Mercury

Just as Solar Physics recognizes that the earth rotates about the sun not in a single orbit, but in many different orbits depending on varying planetary configurations, so also the planet Mercury does not rotate in a single orbit. The effect, however, is more easily observable because the orbits of Mercury are much more elliptical than the earth's and far closer to the sun.

At perihelion the sun's gravitational attraction is more pronounced than the gravitational attraction of the remainder of the solar system. The discrepancy is less pronounced at ahelion. The effect is to advance the orbit in the direction of the sun's rotation at perihelion compared to the effect at ahelion. Mercury's perihelion advances by about 574 arcsec per century (3,600 arcsecs = 1 degree). The situation is illustrated as in the following illustration.

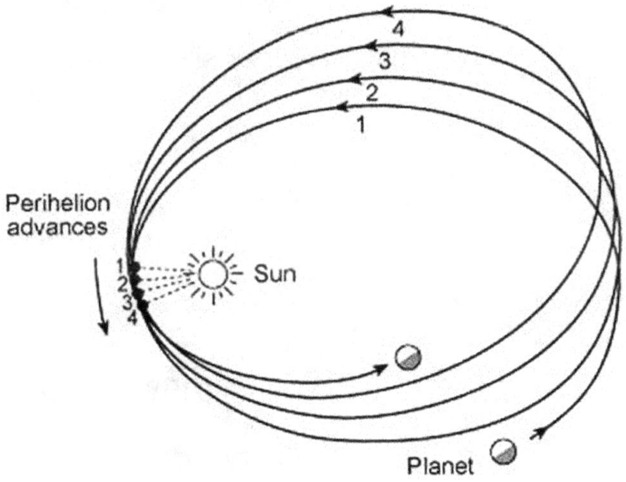

The attempts to explain the observation again illustrate the difference between solar engineering and solar science.

In the scientific view, the orbits of Mercury are what they are. Their observation provides a set of distances and directions from some arbitrary origin. Changes in motion arise from forces which are not observed but must be inferred from observations[1]. For this case Theoretical Physics infers a net force which causes the motion. The vectorial resolution of the net force into other other forces is an open question.

The engineering approach is different. Detailed calculations assuming Newton's law of gravity and the gravitational perturbations of other planets on the motion of Mercury could account for 531 arcsecs. The discrepancy of 43 arcsecs between theory and measurement perplexed astronomers and remained an unresolved issue of the Newtonian theory for many years.

Einstein provided a mathematical answer. He proposed that space and time were warped, under the influence of the sun's gravitation, so as to account for the 43 missing arcsecs.

The argument is nonsense. Neither space nor time are physical realities. Rather they are ideas, like numbers, created for their utility. They are not extended physical objects which can move or be forced. They are merely ideas useful in getting an agreeable answer. It's a case of constructing an engineering model, not of explaining reality.

Tasks (perihelion of Mercury)

Measure the perihelion advance during the upcoming increase in solar activity.

The measurement of Mercury's perihelion is an accumulated effect made over a century. The measurement implies that the effect is the same for

1. Cf *The Second Problem* for a discussion of forces.

each single rotation. Moreover, the accepted calculations imply that Newton's gravitational results have been a constant factor over the century, an implication which is clearly not true.

This investigation would provide

1. evidence of possible variability that is not accounted for.
2. may disprove Einstein's conjecture.
3. suggest some other physical causes not presently considered.
4. reason for reconsidering Newton's gravitational theory.

Having looked at the sun from a coarse understanding, let us now take a closer look.

Let us turn to our main subject: the sun. A good theoretical physicist will first ask: What is the sun? Can we define it?

Definition

A good definition of the sun must be able to identify any particle as included in the definition or not. In this respect it corresponds to a set. A set is truly a set only if one can identify any object as either a member of the set or not. In effect, a definition establishes that we know what we are talking about.

Viewed from the earth the definition of the sun may not seem difficult. It is that object in the sky we can point to. Up close, however, the case may look different.

Consider the earth first. How would we define the earth? Does the earth consist only of solid matter? Or does it also include rivers, lakes, oceans? Does it include the atmosphere? How high? Does it include objects influenced by its gravitation force? Does it include objects influenced by its magnetic force?

As viewed from the sun, the earth appears well-defined as that object far away we could point to. Up close, however, the definition of the earth is not at all apparent.

So also with the sun.

How then does Theoretical Physics proceed?

In *The First Problem*, results are obtained for a *given* measurable set in a defined partition. The measurable set may be a line, a curve, a surface, or a region, material or local.

For the earth, Theoretical Physics would accept any *given* definition of the earth for the object of its analysis.

For the sun, the same.

It is often helpful to consider illustrating the application of an idea of Theoretical Physics to the sun by its analogous application to the earth whenever the application may be appropriate.

We will be using many of the intellectual tools of Theoretical Physics in the following. A synopsis of these is provided in the Appendix.

Task

When making analyses of the sun, define precisely the object of analysis.

Centers of the Sun

In considering the sun a first question arises: What is the center of the sun? Where is the center of the sun?

A mathematical sphere has a center. The sun however, is not a mathematical sphere. Does the sun have a center?

How does Theoretical Physics consider centers?

Centers

Consider a set of physical locations. The physical set is mirrored in the three dimensional set of vectors, **V3**(**r**). Definitions applied to any measurable set define what is under consideration.

Let vectors, **u1R1**, **u2R1**, and **u3R1** be any three non-collinear vectors.

V3(**r**) is partitioned by **pR1** = pr1∗**u1R1** + pr2∗**u2R1** + pr3∗**u3R1**, as explained in *tp1.3*.

Let MR be a measurable set of locations in **V3**(**r**), not necessarily symmetrical. Then for a nested set of partitions based on a partition vector, for any measurable function **f**

I3[MR]∗(**f**|a1; **dR1**) ≡ lim S[MR∩Pn](**f**(Pn(i,j,k))∗**pR1**)/n^3
≡ **sr**(**f**)∗**pR1**
= **I3**[MR](**f**|a1;**dr**)∗**pR1**
for **sr**(**f**) ≡ lim S[MR∩Pn](**f**(Pn(i,j,k)))/n^3.

The proofs of the following theorems are omitted here, but can be examined in *The First Problem*,

Theorem (Center--local)

Given

MR be a measurable set of locations in **V3**(**r**)

a partition vector

pR1 = pr1∗**u1R1** + pr2∗**u2R1** + pr3∗**u3R1**,

m(MR) =(bn[MR∩Pn]/n^3)∗**pR1**

m(MR) = lim S[MR∩Pn](f(Pn(i,j,k)))/n^3

f(**r**,ai) a measurable function at observation ai,
not necessarily continuous over MR,

I3[MR]∗(**f**|ai; **dR1**) = (lim S[bn(MR∩Pn)]
(**f**(Pn(i,j,k)))/n^3)∗**pR1**

sr(**f**) = (lim S[bn(MR∩Pn)](**f**(Pn(i,j,k)))/n^3)

I3[MR]∗(**f**|ai; **dR1**)= **sr**(**f**)∗**pR1**

rf0 = **sr**(**f**)/m(MR).

then

I3[MR]∗((**f**−**rf0**)|ai; **dR1**) = [**0**]

The function **r1**(**x1**,a1) denotes the location of the particle **x1** at observation a1. At different observation ai a different particle may occupy the location **r1**.

So we can contemplate a dual relationship.

Theorem (Center--material)

Given

MX be a measurable set of locations in **V3**(x)

a partition vector

pX1 = pr1∗**u1X1** + pr2∗**u2X1** + pr3∗**u3X1**,

f(**x**,ai) a measurable function at observation ai,
not necessarily continuous over MX,

m(MX) = m(MX)∗**pX1**

I3[MX]∗(**f**|ai; **dX1**) = **sx**(**f**)∗**pX1**

xf0 = **sx**(**f**)/m(MX).

then

$$\mathbf{I3}[MX]*((\mathbf{f}-\mathbf{xf0})|ai;\ \mathbf{dX1}) = [\mathbf{0}]$$

It follows that invergences

$$\mathbf{I3}[MR]\bullet((\mathbf{f}-\mathbf{rf0})|ai;\ \mathbf{dR1}) = 0$$

$$\mathbf{I3}[MX]\bullet((\mathbf{f}-\mathbf{xf0})|ai;\ \mathbf{dX1}) = 0$$

and incurls

$$\mathbf{I3}[MR]\wedge((\mathbf{f}-\mathbf{rf0})|ai;\ \mathbf{dR1}) = \mathbf{0}$$

$$\mathbf{I3}[MX]\wedge((\mathbf{f}-\mathbf{xf0})|ai;\ \mathbf{dX1}) = \mathbf{0}$$

equal zero as well.

The vectors **rf0** and **xf0** are called **centers of f over MR and MX** respectively.

Also for some any fixed value of the function

Given

rf1 a fixed value of **f**(**r**);

xf1 a fixed value of **f**(**x**);

then

$$\mathbf{I3}[MR]*(\mathbf{f}-\mathbf{rf1})|ai;\ \mathbf{dR1} = (\mathbf{rf0}-\mathbf{rf1})*\mathbf{m}(MR)$$

$$\mathbf{I3}[MX]*(\mathbf{f}-\mathbf{xf1})|ai;\ \mathbf{dX1} = (\mathbf{xf0}-\mathbf{xf1})*\mathbf{m}(MX)$$

The centers of a function clearly depend on the function and may be located anywhere in MR or MX.

The next question is: can we find a center of symmetry?

Theorem (Center of Symmetry)

Given

MR a measurable set in **V3(r)**

with measure m(MR)***pR1**;

a partition vector

$$\mathbf{pR1} = pr1*\mathbf{u1R1} + pr2*\mathbf{u2R1} + pr3*\mathbf{u3R1};$$

f a measurable scalar function over MR;

I3[MR]∗((f−rf0)|ai;**dR1**) = **0**;

sr(f) = **I3**[MR](f(**r**)|ai;**dR1**);

sr(f∗**r**) = **I3**[MR](f(**r**)∗**r**|ai;**dR1**);

sr0(f) = **sr**(f∗**r**)/sr(f);

and dually

MX a measurable set in V3(**x**)

with measure m(MX) ∗**pX1**

a partition vector

pX1 = pr1∗**u1X1** + pr2∗**u2X1** + pr3∗**u3X1;**

f a measurable scalar function over MX

I3[MX]∗((f−xf0)|ai; **dX1**) = **0**

sx(f) = **I3**[MX](f(**x**)|ai; **dX1**)

sx(f∗**x**) = **I3**[MX](f(**x**)∗**x**|ai; **dX1**)

sx0(f) = **sx**(f∗**x**)/sx(f)

then

I3[MR] ∗((f∗(**r**−**sr0**(f)))|ai; **dR1**) = **[0]**

I3[MX] ∗((f∗(**x**−**sx0**(f)))|ai; **dX1**) = **[0]**

Moreover,

sr0(f) = **sr**(f∗**r**)/(rf0∗m(MR))

sx0(f) = **sx**(f∗**x**)/(xf0∗m(MX))

The vectors **sr0(**f) and **sx0(**f) are called centers of local symmetry and centers of material symmetry.

Note that the function needs to be a scalar function..

The physical units of **sr0** and **sx0** are L. (extension).

There's more.

Theorem (Centroids)

Given

MR a measurable set in V3(**r**)

with measure m(MR) ∗**pR1**

a partition vector

pR1 = pr1∗**u1R1** + pr2∗**u2R1** + pr3∗**u3R1**;

f a measurable scalar function

of neutral physical units over MR

cr0(f) = **sr**(f∗**r**)/m(MR)

and dually

MX a measurable set in V3(**r**)

with measure m(MX) ∗**px**

a partition vector

pX1 = pr1∗**u1X1** + pr2∗**u2X1** + pr3∗**u3X1**;

f a measurable scalar function

of neutral physical units over MX

cx0(f) = **sx**(f∗**x**)/m(MX)

Then

I3[MR] ∗((f∗**r**−**cr0**(f))|ai; **dR1**) = [**0**]

I3[MX] ∗((f∗**x**−**cx0**(f))|ai; **dX1**) = [**0**]

The vectors **cr0**(f) and **cx0**(f) are called the **local and material centroids of f over MR and MX** respectively. The physical units of **cr0** and **cx0** are L.

Centroids are related to centers of symmetry as.

cr0(f) = rf0∗**sr0**(f)

cx0(f) = xf0∗**sx0**(f).

Theoretical Physics contemplates, in general, a large number of different “centers” suitable for analyzing the sun.

Specific results come from specifying specific functions. So let us consider the density of a particle generally so that afterwards we may apply our knowledge to the sun.

Let

r(**x**,a1) describe the position of particles in the material universe at observation *a1*;

r1(**x1**,a1) be the position of particle **x1** at observation a1;

EX be an open subset of V3(**x**), physical matter, containing **x1** in the universe at rest;

X1, a subset of EX, be a rectangular parallelepiped with sides

x1 + hx1∗**u1** - **x1** = hx1∗**u1,**

x1 + hx2∗**u2** - **x1** = hx1∗**u1,**

and

x1 + hx3∗**u3** - **x1** = hx1∗**u1**

for some hxi>0.

r(EX) ≡ EXR

r(X1,a1) ≡ RX1 be the image of X1 under **r**(**x**,a1) which must be a subset of EXR which consequently must contain **r1**.

The material occupying RX1 is appropriately described by **x**(RX1,a1) = X1.

Since **r**(**x**) is continuous over EX, EXR will also be open.

An illustration of these entities may prove helpful.

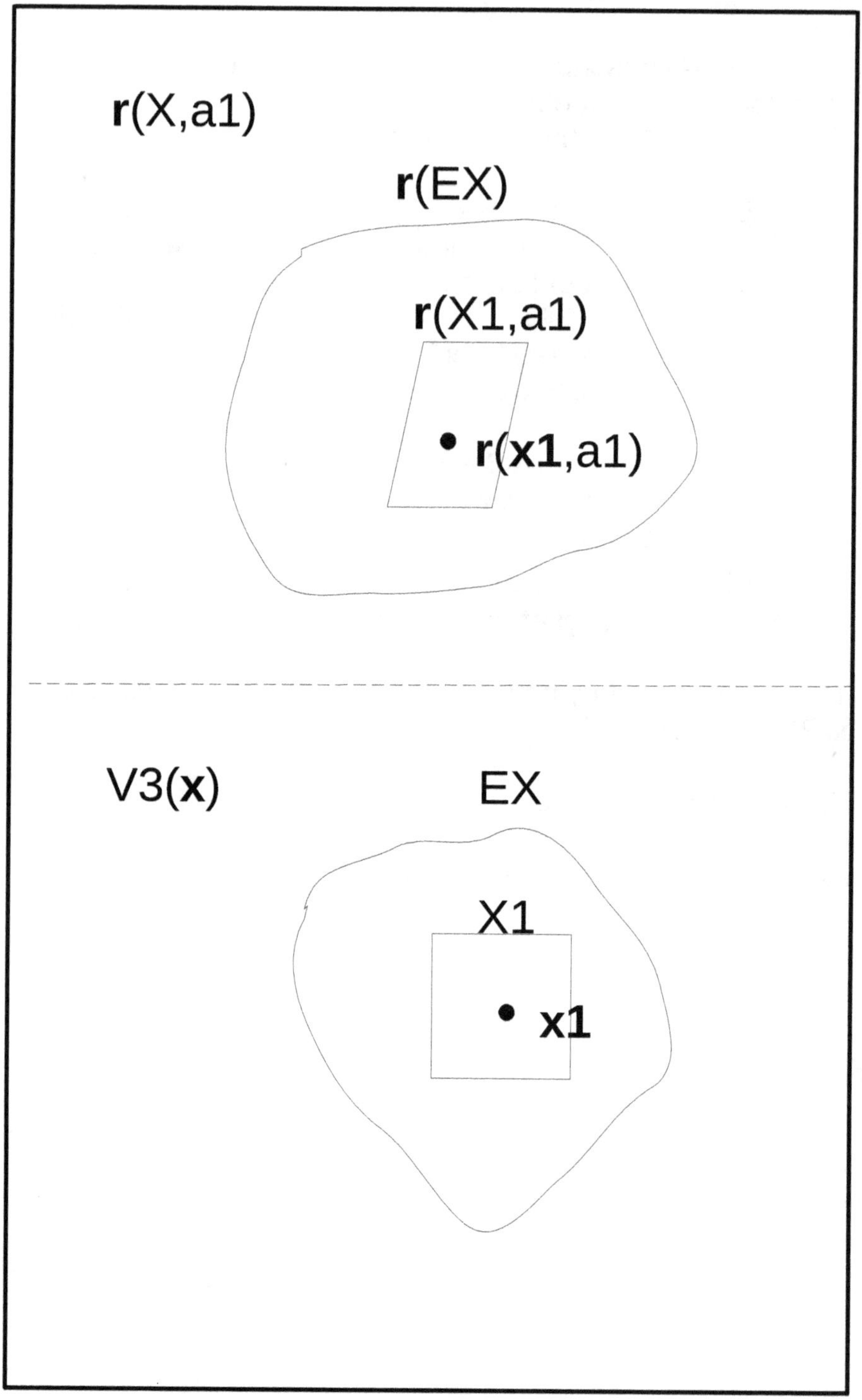
r(X,a1)
r(EX)
r(X1,a1)
r(**x1**,a1)
V3(**x**)
EX
X1
x1

Dually let

x(r,a1) be the dual description of r(x,a1);

ER be an open subset of V3(r) containing r1 in the universe at rest;

R1, a subset of ER, be a parallelopiped with sides

r1 + hr1∗u1 - r1,

r1 + hr2∗u2 - r1,

and

r1 + hr3∗u3 - r1;

x(ER) ≡ ERX

x(R1,a1) ≡ XR1 be the image of R1 under x(r,a1) which must be a subset of ERX and which consequently must contain **x1**. The space occupied by XR1 is appropriately described by **r**(XR1,a1) = R1.

If **x** is continuous over ER, ERX will also be open.

For any open EX and ER, each of the above subsets has a non-zero volume, symbolized by vol(X1), vol(R1), vol(XR1), and vol(RX1). Consequently ratios of these volumes may be attached to the particle **x1** occupying location **r1** or dually to the location **r1** occupied by **x1** at observation *a1*.

Consider now the gradient **D3[x1]**∗(**r**|a1; **dx**). The determinant of the gradient is a kind of volume. Let us determine if the determinant of the gradient has a relationship to the above volumes.

det[**D3[x1]**∗(**r**|a1;**dx**)]

= lim det[(**r**(**x1**+hx1∗**u1**,a1) - **r1**)∗**u1**/hx1

+ (**r**(**x1**+hx2∗**u2**,a1) - **r1**)∗**u2**/hx2

+ (**r**(**x1**+hx3∗**u2**,a1) - **r1**)∗**u3**/hx3]

= (**r**(**x1**+hx1∗**u1**,a1)- **r1**)∧(**r**(**x1**+hx2∗**u2**,a1)- **r1**)

•(**r**(**x1**+hx3∗**u2**,a1)- **r1**)

/(hx1∗hx2∗hx3).

The denominator is the material volume the first octant of the set X1 defined by vectors hx1∗**u1**, hx2∗**u2**, hx3∗**u3**. The numerator is the volume of the local image of the octant of X1, in RX1.

Consequently det[**D3[x1]**∗(**r**|a1;**dx**)] can be thought of as the ratio of two volumes. So we do well to consider gradients carefully.

Theoretical Physics contemplates many different gradients. Here let us consider only the most restrictive kind, namely simply continuous gradients where **r**(**x**) and **x**(**r**) are locally differentiable at **r1**(**x1**,a1) with gradients having octants for both domains and ranges. Then[1]

- *material and local octant gradients exist and may be succinctly described by*
  D3[r1]∗*(**r**(**x**,a)|a1;**dx**) or by*
  D3[x1]∗*(**x**(**r**,a)|a1;**dr**);*
- **D3[r1]**∗(**r**(**x**,a)|a1;**dx**) and **D3[x1]**∗(**x**(**r**,a)|a1;**dr**) are inverses of each other;

Such gradients over all eight octants are equal. Their determinants in combination is the numerator is the local volume of a set symbolized above as RX1.

Consequently for this restrictive case,
vo(RX)/vol(X1)
= 8∗(**r**(**x1**+hx1∗**u1**,a1)− **r1**)^(**r**(**x1**+hx2∗**u2**,a1)− **r1**)
•(**r**(**x1**+hx3∗**u2**,a1)− **r1**)
/8∗(hx1∗hx2∗hx3)
⟶ det[**D3[x1]**∗(**r**|a1; **dx**)]

Thus the determinant of the first quadrant gradient may be interpreted as linear map whose range is vol(X1) with image vol(RX1), that is

vol(X1)∗det[**D3[x1]**∗(**r**|a;**dx**)] = vol(RX1)

---

1. See tp1.4

We then define the **relative-size** of **x1** at location **r1** at observation a1 as

rsz(**x1**,a1) ≡ det[**D3[x1]**∗(**r**|a1; **dx**)]

Likewise, we defne the **relative-density** at location **r1** of the particle **x1** for observation *a1*.

rdn(**r1**,a1) ≡ det[**D3[r1]**∗(**x**|a1;**dr**)]

These ideas correspond to our observations that when more matter is squeezed into a given volume, the density increases. Similarly when a given quantity of matter is expanded into a larger volume the density decreases.

The physical units of the relative-size of a material particle are

L∗L∗L/(L∗L∗L);

those of the associated relative-density are similarly

L∗L∗L/(L∗L∗L).

In general rdn and rsz are tied to gradients of various types.

We list here a significant theorem in this context to illustrate the kind of intellectual tools available to Theoretical Physics.

Theorem (expansion)

Given

r(**x,**a1), an observation of the universe at a1
r1(**x1**,a1) the reference;
Rj and XRj mutually related sections
at **r1**(**x1**,a1) under **r**(**x**,a1);
D3[x1]∗(**r**|a1;**dXRj**) and **D3[r1]**∗(**x**|a1;**dRj**),
mutually related gradients;
conditions for which the D**D3** interchange holds

for

n consistently denoting forward or backward ;
tr the trace function of a matrix;
det the determinant function of a matrix;
rsz the relative size of **x1**(**r1**,a1);
rdn the relative density of the particles
at **r1**(**x1**,a1);
y=x;
Rj and XRj observationally stable
over the interval [a1−da,a1+da]
D[a1](**r**|**x1**;d_fa) directed from **r1**(**x1**,a1)
into section Rj;
D[a1](**r**|**x1**;d_ba) directed from section Rj
to **r1**(**x1**,a1)

then

tr[**D3[r1]**∗(D[a1](**r**|**x1**;d_na)|a1;**dRj**)•**[URj**$^{-1}$**]•T[URj**$^{-1}$**]**]
= D[a1](det[**D3[x1]**∗(**r**|a1;**dXRj**)]|**x1**;d_na)
/det[**D3[x1]**∗(**r**|a1;**dXRj**)]
= D[a1](rsz(XRj,a)|**x1**;d_na)/rsz(XRj,a1)
= −D[a1](det[**D3[r1]**∗(**x**|a1;**dRj**)]|**x1**;d_na)
/det[**D3[r1]**∗(**x**|a1;**dRj**)]
= −D[a1](rdn(Rj,a)|**x1**;d_na)/rdn(Rj,a1)

Dually,

given

x1(**r1**,a1) the reference of interest;
Xj and RXj mutually related sections
at **x1**(**r1**,a1) under **x**(**x**,a1);

D3[r1]∗(**x**|a1;**dRXj**) and **D3[x1]**∗(**r**|a1;**dXj**),
mutually related gradients;

for

Xj and RXj be observationally stable
over the interval [a1−da,a1+da]

D[a1](**x**|**r1**;d_fa) directed from **x1**(**r1**,a1)
into section Xj;

D[a1](**x**|**r1**;d_ba) directed from section Xj
to **x1**(**r1**,a1)

then

tr[**D3[x1]**∗(D[a1](**x**|**r1**;d_na)|a1;**dXj**)•**[UXj^{-1}]•T[UXj^{-1}]**]
= D[a1](det[**D3[r1]**∗(**x**|a1;**dRXj**)]|**r1**;d_na)
/det[**D3[r1]**∗(**x**|a1;**dRXj**)]
= D[a1](rdn(RXj,a)|**r1**;d_na)/rdn(RXj,a1)
= −D[a1](det[**D3[x1]**∗(**r**|a1;**dXj**)]|**r1**;d_na)
/det[**D3[x1]**∗(**r**|a1;**dXj**)]
= −D[a1](rsz(Xj,a)|**r1**;d_na)/rsz(Xj,a1)

From this we learn that an analysis of the sun will need to examine gradients carefully.[1]

Now local relative density[2], rdn(**r**,ai), is a scalar function with neutral units defined over MR while material relative size, rsz(**x**,ai), is a scalar function with neutral units defined over MRX.

Theorem (centers of density)

Given

MR a measurable set in V3(**r**)
with measure m(MR)∗**pR1**;

a partition vector
pR1 = pr1∗**u1R1** + pr2∗**u2R1** + pr3∗**u3R1;**

1. A synopsis of derivatives in general and gradients in particular is given in the Appendix
2. This significant function is discussed amply in *The Second Problem,*

rrdn0 the center of rdn(**r**) over MR;
rr0 the center of **r** over MR;
rx0 the center of **x**(**r**) over MR;
sr0(rdn) the center of symmetry of rdn(**r**) over MR;
cr0(rdn) the centroid of rdn(**r**) over MR;

Dually

MRX = {**x**(**r**,ai)|**r** in MR};
pX1 = px1∗**u1X1** + px2∗**u2X1** + px3∗**u3X1**,
a partition vector of V3(**x**);
xrsz0 the center of rsz(**x**) over MRX;
xx0 the center of **x** over MRX;
xr0 the center of **r**(**x**) over MRX;
sx0(rsz) the center of symmetry of rsz(**x**) over MRX;
cx0(rsz) the centroid of rsz(**x**) over MRX;

then

rrdn0 = 1/xrsz0 = m(MRX)/m(MR)

I3[MR]∗(**r**|ai; **dR1**) ≡ m(MR)∗ **rr0**∗**pR1**
= m(MR)∗ **sr0**(1) ∗**pR1**
= m(MR)∗ **cr0**(1) ∗**pR1**
= xrsz0∗m(MRX)∗ **rr0**∗**pR1**

I3[MR]∗(**x**|ai; **dR1**) ≡ m(MR)∗ **rx0**∗**pR1**
= xrsz0∗m(MRX)∗ **rx0**∗**pR1**
= xrsz0∗m(MRX)∗ **sx0**(rsz) ∗**pR1**
= m(MRX)∗ **cx0**(rsz) ∗**pR1**

I3[MRX]∗(**x**|ai; **dX1**) ≡ m(MRX)∗ **xx0**∗**pX1**
= m(MRX)∗ **sx0**(1) ∗**pX1**
= m(MRX)∗ **cx0**(1) ∗**pX1**
= rdn0∗m(MR) ∗**xx0**∗**pX1**

I3[MRX](**r**|ai; **dX1**) ≡ m(MRX)∗ **xr0**∗**pX1**
= rdn0∗m(MR)∗ **xr0**∗**pX1**
= rdn0∗m(MR)∗ **sr0**(rdn) ∗**pX1**
= m(MR)∗ **cr0**(rdn) ∗**pX1**

We could write corresponding dual results.

Here are results for some specific functions.

Corollary

$$\mathbf{rr0} = \mathbf{sr0}(1) = \mathbf{cr0}(1)$$
$$\mathbf{rx0} = \mathbf{sx0}(\mathrm{rsz})$$
$$\mathbf{xx0} = \mathbf{sx0}(1) = \mathbf{cx0}(1)$$
$$\mathbf{xr0} = \mathbf{sr0}(\mathrm{rdn})$$

$$\mathbf{cr0}(\mathrm{rdn})\bullet\mathbf{cx0}(\mathrm{rsz}) = \mathbf{sr0}(\mathrm{rdn})\bullet\mathbf{sx0}(\mathrm{rsz})$$
$$\mathbf{cx0}(\mathrm{rsz})\wedge\mathbf{cr0}(\mathrm{rdn}) = \mathbf{sx0}(\mathrm{rsz})\wedge\mathbf{sr0}(\mathrm{rdn})$$
$$\mathbf{cr0}(\mathrm{rdn})*\mathbf{cx0}(\mathrm{rsz}) = \mathbf{sr0}(\mathrm{rdn})*\mathbf{sx0}(\mathrm{rsz})$$
$$\mathbf{cx0}(\mathrm{rsz})\bullet\mathbf{cr0}(\mathrm{rdn}) = \mathbf{sx0}(\mathrm{rsz})\bullet\mathbf{sr0}(\mathrm{rdn})$$
$$\mathbf{cr0}(\mathrm{rdn})\wedge\mathbf{cx0}(\mathrm{rsz}) = \mathbf{sr0}(\mathrm{rdn})\wedge\mathbf{sx0}(\mathrm{rsz})$$
$$\mathbf{cx0}(\mathrm{rsz})*\mathbf{cr0}(\mathrm{rdn}) = \mathbf{sx0}(\mathrm{rsz})*\mathbf{sr0}(\mathrm{rdn})$$

The following explains how rdn and rsz combine with other functions.

Theorem (centers of relative density and size functions)

Given

MR a measurable set in V3(**r**) with measure m(MR)***pR1**;

a partition vector

$$\mathbf{pR1} = \mathrm{pr1}*\mathbf{u1R1} + \mathrm{pr2}*\mathbf{u2R1} + \mathrm{pr3}*\mathbf{u3R1};$$

rdn0 the center of rdn(**r**) over MR;

f a vector function over MR;

rf0 the center of **f** over MR;

MRX = {**x**(**r**,ai)|**r** in MR};

rsz0 the center of rsz(**x**) over MRX;

a partition vector

$$\mathbf{pX1} = \mathrm{px1}*\mathbf{u1X1} + \mathrm{px2}*\mathbf{u2X1} + \mathrm{px3}*\mathbf{u3X1},$$

xf0 the center of **f** over MRX;

then

$$\mathbf{I3}[\mathrm{MR}]*((\mathrm{rdn}(\mathbf{r})*\mathbf{f})|\mathrm{a1};\ \mathbf{dR1}) = \mathrm{rdn0}*\mathbf{xf0}*\mathrm{m(MR)}*\mathbf{pR1}$$
$$\mathbf{I3}[\mathrm{MRX}]*((\mathrm{rsz}(\mathbf{x})*\mathbf{f})|\mathrm{a1};\ \mathbf{dX1}) = \mathrm{rsz0}*\mathbf{rf0}*\mathrm{m(MRX)}*\mathbf{pX1}$$

If rdn(**r**) is constant over MR or rsz(**x**) is constant over MRX,
then

xf0 = **rf0**

Theorem (rectangular partitions)
Given
MR a measurable set in V3(**r**)
with measure m(MR)∗**pr**;
a partition vector
pr = pr1∗**u1** + pr2∗**u2** + pr3∗**u3**,
;
f a scalar function over MR;
rf0 the center of f over MR;
sr0(f) the center of symmetry of f over MR;

MRX = {**x**(**r**,ai)|**r** in MR};
a partition vector
px = px1∗**u1** + px2∗**u2** + px3∗**u3**;
xf0 the center of f over MRX;
sx0(f) the center of symmetry of f over MRX;
then

xf0∗**I3**[MR]((f(**r**)∗**r**)|ai; **dr**)
= rf0∗rsz0∗**I**3[MRX]((f(**x**)∗**x**)|ai; **dx**)
•**G**(**q**(**sx0**(f))) •**G**(**sr0**(f))

rf0∗**I3**[MRX]((f(**x**)∗**x**)|ai; **dx**)
= xf0∗rdn0∗**I**3[MR]((f(**r**)∗**r**)|ai; **dr**)
•**G**(**q**(**sr0**(f))) •**G**(**sx0**(f))

where **q** is the quotient vector and **G** is the diagonal vector operator.

Further for f with neutral units.

I3[MR]((f(**r**)∗**r**)|ai;**dr**)
= rsz0∗**I3**[MRX]((f(**x**)∗**x**)|ai;**dx**)
•**G**(**q**(**cx0**(f)))•**G**(**cr0**(f))

I3[MRX]((f(**x**)∗**x**)|ai;**dx**)

= rdn0∗**I3**[MR]((f(**r**)∗**r**)|ai;**dr**)

•G(**q**(**cr0**(f)))•**G**(**cx0**(f))

Corollary

If **rr0** is the center of **r** over MR,

I3[MR]^((**r**−**rr0**)|ai;**dR1**) = **0**

I3[MR]•((**r**−**rr0**)|ai;**dR1**) = 0

If **xx0** is the center of **x** over MX,

I3[MX]^((**x**−**xx0**)|ai;**dX1**) = **0**

I3[MX]•((**x**−**xx0**)|ai; **dX1**) = 0

We might also ask how these results are affected by translations of the underlying set.

Theorem (translation)

Given

MR a measurable set in V3(**r**)
with measure m(MR)∗**pr**;

pz = p1∗**u1** + p2∗**u2** + p3∗**u3**, a partition vector

pr = p1∗**u1** + p2∗**u2** + p3∗**u3**, a partition vector

A a fixed transformation of rank 3

b a constant vector

MRZ = {**z**|**z** = **r**•**A** + **b**, **r** in MR}

f(**z**(**r**)) a measurable compound vector function over MR;

I3[MR]∗((**f**−**rf0**|ai;**dr**) = **0**;

f(**z**(**r**)) a measurable compound scalar function over MR;

I3[MR]∗((f−rf0)|ai; **dr**) = **0**;

sr0(f) = **sr**(f∗**r**)/sr(f);

cr0(f) = **sr**(f∗**r**)/m(MR);

sz0(f) = **sr0**(f)•**A** + **b**

cz0(f) = **cr0**(f)•**A** + **b**

or dually

MX a measurable set in V3(**x**) with measure m(MX) ∗**px**;

px = p1∗**u1** + p2∗**u2** + p3∗**u3**, a partition vector

MXZ = {**z**|**z** = **r•A** + **b**, **r** in MX}

f(**z**(**r**)) a measurable compound vector function over MX;

I3[MX]∗((**f**−**xf0**|ai;**dx**) = **0**;

f(**z**(**x**)) a measurable compound scalar function over MX;

I3[MX]∗((f−xf0)|ai;**dx**) = **0**;

sx0(f) = **sx**(f∗**x**)/sx(f);

cx0(f) = **sx**(f∗**x**)/m(MX);

sz0(f) = **sx0**(f)**•A** + **b**

cz0(f) = **cx0**(f)**•A** + **b**

then

for **f** a vectorial function

I3[MRZ]∗((**f**(**z**)−**rf0**)|ai;**dz**) = **0**

I3[MXZ]∗((**f**(**z**)−**xf0**)|ai;**dz**) = **0**

for f a scalar function

I3[MRZ]∗(f(**z**)∗(**z**−**sz0**(f))|ai;**dz**) = **0**

I3[MXZ]∗(f(**z**)∗(**z**−**sz0**(f))|ai; **dz**)= **0**

and for f with neutral units

I3[MRZ]∗((f(**z**)∗**z**−**cz0**(f))|ai;**dz**) = **0**

I3[MXZ]∗((f(**z**)∗**z**−**cz0**(f))|ai;**dz**) = **0**

Consequently, while the center of symmetry and centroid of f over MR are dilated whenever MR is linearly dilated into MRZ, the center of **f** remains unchanged over both sets.

Corollary

Given

For the linear dilation and translation

$$\mathbf{z}(\mathbf{r}) = \mathbf{r} \bullet \mathbf{A} + \mathbf{b}$$

or dually

$$\mathbf{z}(\mathbf{x}) = \mathbf{x} \bullet \mathbf{A} + \mathbf{b}$$

let

rz0 = **rr0•A** + **b**
sz0(rdn) = **sr0**(rdn)**•A** + **b**
cz0(rdn) = **cr0**(rdn)**•A** + **b**
xz0 = **xr0•A** + **b**
sz0(rsz) = **sx0**(rsz)**•A** + **b**
cz0(rsz) = **cx0**(rszn)**•A** + **b**

then

rz0 = **sz0**(rdn) = **cz0**(rdn)
xz0 = **sz0**(rsz) = **cz0**(rsz)

For a measurable set Theoretical Physics conceives many ideas of "center" which can be used to describe of analyze the sun. The following table summarizes these.

Symbol	Name	definition	units
sr(**f**)	The scalar part of **I3**	**I3**[MR]∗(**f**|ai; **dR1**) = **sr**(**f**)∗**pR1**	u(f)
rf0	Center of f in MR	**sr**(**f**)/m(MR)	u(f)
sr(f∗**r**)	Scalar part of the integral	lim S[bn(MRÇPn)] (f(Pn(i,j,k))∗**r**(Pn(i,j,k)/n^3)	u(f)∗**L**
sr0(f)	Center of symmetry	**sr**(f∗**r**)/sr(f)	L
cr0(f)	Centroids	**sr**(f∗**r**)/m(MR)	u(f)∗**L**
rrdn0	Center of density	f = rdn(**r**)	neutral

Symbol	**Name**	**definition**	**units**
sx(**f**)	The scalar part **I3**	**I3**[MRX]∗(**f**|ai; **dX1**) = **sx**(**f**)∗**pX1**	u(f)
xf0	Center of f in MX	**sx**(**f**)/m(MRX)	u(f)
sx(f∗**x**)	Scalar part of the integral	lim S[bn(MRX**Ç**Pn)] (f(Pn(i,j,k))∗**x**(Pn(i,j,k)/n^3)	u(f)∗**L**
sx0(f)	Center of symmetry	**sx**(f∗**x**)/sx(f)	L
cx0(f)	Centroids	**sx**(f∗**x**)/m(MRX)	u(f)∗**L**
xrsz0	Center of size	f = rsz(**x**)	neutral

So just how are these results applied to the sun?

The centers defined above are all results of integrals over defined measurable sets. Since this is so, centers over non-intersecting sets may be added. That is,
for MR1∩MR2 = null

rf0(MR1∪MR2) = **rf0**(MR1) + **rf0**(MR2)

This hold true for all centers as well: **sr0**(f), **cr0**(f), rrdn0 and their duals.

Consequently, the set MR, identified as the sun, may be intellectually partitioned, coarsely or finely as may be desired, for analysis.

By analogy to the earth, one might seek to analyze the earth first as solid matter, then add the oceans, and further add the atmosphere. Centers for each may be analyzed separately and later combined.

Thus a large prospect for analysis opens.

Since the centers of symmetry designate a definite location or particle, they would appear especially suitable for analysis.

Relative densities and relative sizes also invite a thorough analysis. In this respect a closer inspection of kinds of gradients the sun supports is imperative.

We will do well also to focus on the observables of the sun, especially its surface and its atmosphere.

As illustration let us try a simple and crude analysis.

Analysis

Let us start with the analysis of the photograph of granules produced by the High-resolution image of the Sun's surface taken with the Daniel K. Inouye Solar Telescope (DKIST). on January 20, 2020.

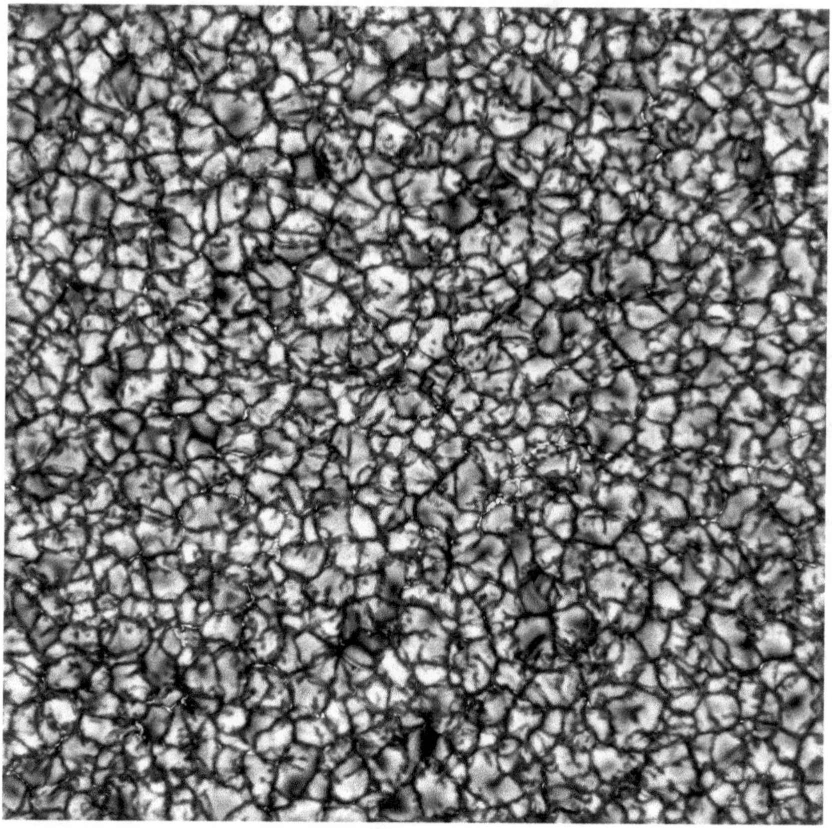

This image at the time of publication was the highest resolution image of the sun's surface ever taken. The image is filtered at 789 nanometers (nm), a frequency in the ultraviolet just a bit higher than the optical spectrum.

.

The image covers an area 36,500 × 36,500 km. The earth in comparison has a mean radius of 6371 km.

Features as small as 30 km (18 miles) in size are observable. The image shows a pattern of turbulence of the cell-like structures, each about the size of Texas (approximately 700,000 km^2). The inter-cellular areas appear as dark boundaries or lanes within which bright streaks often appear.

For our analysis consider only a single cell.

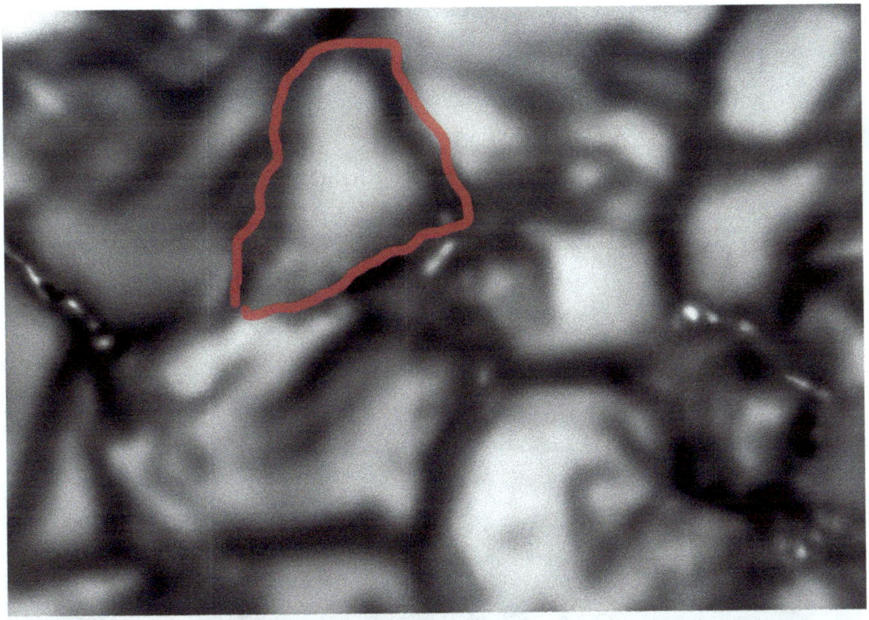

The image covers about 4000 km by 4000km

The structure to be analyzed is roughly equilateral with side of 700 km (area 850 km^2) by 230 km in depth or about 200,000 km^3 of the sun.

What is its measure?

m(M|**pV1**) = vol(M)/vol(P1(i,j,k)).
vol(M) = 200,000
m(M|**pV1**) = 200,000/1
which assumes a partition vector of **p** ≡ **u1** **+u2** **+u3**, that is a 1 km cube.

Now we can calculate its center of symmetry
sr0(1) =**sr**(**r**)/sr(1)
sr(1) is just m(M|**p1**) = 200,000
sr(**r**) = lim S[bn(MR∩Pn)](**r**(Pn(i,j,k)/n^3)

Let

i run from 0 to 606
j from -350 to 350
k from 1 to 230

Then **sr**(**r**) = (303,0,115), that is 115 km below the center of the visual triangle as shown in the following image.

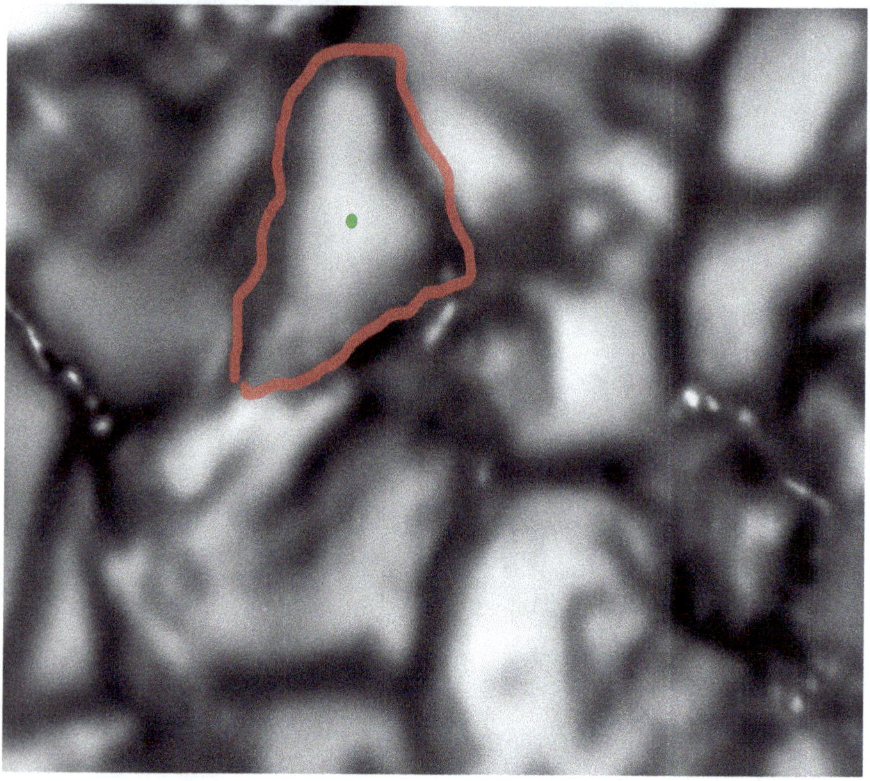

This simple and crude illustration shows how a center symmetry may be calculated for a small region of the sun.

The analysis clearly depends on what is observable.

Many other centers await analysis.

Tasks

Analyze the sun's

centers of symmetry;
centers of relative density and relative size;
what kind of gradients it supports.

Surfaces

Theoretical Physics supplies some ideas about surfaces and, related to surfaces, of step functions.

What is immediately observable of both the sun and the earth are their surfaces.

What is a surface?

The theoretical physicist defines a surface as follows:

Definition (Surface of a measurable set)

Given

M, a measurable set of **V3**;

pR1 = (p1∗**u1R1**+p2∗**u2R1**+p3∗**u3R1**);

Pn, a sequenced set of partitions of **V3** based on **pR1**;

then

if Pn(i,j,k) ∩ M = Pn(i,j,k)

Pn(i,j,k) is called a **cell interior to** M

if Pn(i,j,k) ∩ M = null

Pn(i,j,k) is called a **cell exterior to** M

if Pn(i,j,k) ∩ M ⊂ Pn(i,j,k) properly

Pn(i,j,k) is called a **surface cell of** M.

end of definition

The definition can be applied to any measurable set and so defines what is under consideration. The physical set is mirrored in the three dimensional set of vectors, **V3**. Let vectors, **u1R1**, **u2R1**, and **u3R1** be any three non-collinear vectors. **V3** is then partitioned by **pR1** as explained in *tp1.3:* Then each individual cell in each of the sequence of partitions is interrogated to establish whether or not it lies exterior to M, interior to M, or partly in both.

The set
SM ≡ {lim **r**(Pn(i,j,k))|Pn(i,j,k) is a surface cell} is called the **surface of** M.

In this connection, it is well to understand integration in Theoretical Physics.

INTEGRALS OF THEORETICAL PHYSICS

Theoretical Physics conceives an integral to match each of its derivatives. For each it defines a measure, the conditions for interior cancellation, and the integration of impulse functions. From one viewpoint, these new integrals are extensions of the fundamental theorem of integral calculus; in Theoretical Physics they also are seen as defining an extended class of functions suitable for describing, analyzing and explaining physical reality.

Observational Integration

Definition (observational integrals)

Given

{**r**(**x**,a)} the set of ordered observations;
f(**r**(**x**,a)), a generalized function of the observations;
Pn a set of partitions of the observations
based on the partition scalar pt;
m(Pni), the measure of a cell in the Pn partition;
M a measurable set of the observations;

for

bn(M∩Pn) the number of cells of Pn intersecting M;
S[bn(M∩Pn)], a summation over cells indicated by bn;

then

I[M](**f**(**r**(**x**,a));da)
≡ lim S[bn(M∩Pn)](**f**(a(Pni))∗m(Pni))
= pt∗lim S[bn(M∩Pn)](**f**(a(Pni))/n) as n→∞.

end of definition

When M is an interval [a0,an], the integral is written as
I[a0,an](**f**(**r**(**x**,a));da)

≡ lim S[bn(Pn∩[a0,an])](**f**(**r**(**x**,a0+i*da))*da)

where

a(Pni) = a0 + i*pt/n
da ≡ pt/n
= a(Pn(i+1))−a(Pni).

In Theoretical Physics integration with respect to the observational index may be applied either generically or specifically as for example,

Specifically: **I**[ ](**f**(**r**(**x1**,a))|**x1**;da)
Genericallly: **I**[ ](**f**(**r**(**x**,a))|**x**;da).

In the generic case, **r**(**x**,a) represents continuous observations of locations of material particles throughout the whole universe, or more modestly, over a measurable region of physical locations. Observational integration yields specific results in the first case or results for the region in the second.

A local trace, **r**(**x1**,a) denotes the location of a material particle **x1** at observation a. Specific observational integration yields a result for particle **x1**. Such integrals are symbolized

I[ ](**f**|**x1**;da) ≡ **I**[ ](**f**(**r**(**x1**,a));da).

Likewise material tracks, **x**(**r1**,a) represent observations of the particles occupying location **r1**. Specific observational integration yields a result for location **r1**. Such integrals may be symbolized

I[ ](**f**|**r1**;da) ≡ **I**[ ](**f**(**x**(**r1**,a));da).

For dealing with surfaces, step functions are helpful. Theoretical Physics provides a variety of them

Step Functions

Consult tp1.4 for an extensive discussion of these functions. Only two are given here.

Definition (observational step functions)

Given

{**r**(**x**,a)}, the set of ordered observations;

for

a1, a reference observation

$$\text{ua}(\mathbf{r}(\mathbf{x},a)-\mathbf{r}(\mathbf{x},a1)) \equiv 1,\ a>a1$$
$$\equiv 0\ \ a\leq a1$$

and

$$\text{va}(\mathbf{r}(\mathbf{x},a)-\mathbf{r}(\mathbf{x},a1)) \equiv 1,\ a\geq a1$$
$$\equiv 0\ \ a<a1$$

are the **unit observational step functions**.

end of definition

The unit observational step functions, may be used to designate the start of an interval of observations. The functions 1-ua or 1-va may then be used to designate the end of the interval.

Another useful step function relates to regions.

Definition (regional step functions in V3)

Given

r(**x**,a1), a single observation;

MX, a measurable set of particles in V3(**x**);

SMX, the surface of MX;

then

$$\text{ux}(\mathbf{x},a1)|\text{MX} \equiv 0,\quad \mathbf{x} \text{ in MX}\cup\text{SMX}$$
$$\equiv 1,\quad \text{otherwise}$$
$$\text{vx}(\mathbf{x},a1)|\text{MX} \equiv 0,\quad \mathbf{x} \text{ in MX}-\text{SMX}$$
$$\equiv 1,\quad \text{otherwise.}$$

Dually,
given
MR, a measurable set of locations in V3(**r**);
SMR, the surface of MR;
then

ur(**r**,a1)|MR ≡ 0, **r** in MR∪SMR
≡ 1, otherwise
vr(**r**,a1)|MR ≡ 0, **r** in MR–SMR
≡ 1, otherwise.
end of definition

Step functions may also be differentiated in which case they yield a variety of impulse functions.

The scope of these ideas can be far-ranging.

At a single observation a1, a function may be represented almost everywhere as a combination of a continuous function and combinations of step functions.

f(**x**(**r**,a1))
= **f1**(**x**(**r**,a1))
+ S[1,n1](**I3**[SECT(**x1**,**u1XS1i**,**u2XS1i**,**u3XS1i**)]
∗(ux(**x1**|SECT(**x1**,**u1XS1i**,**u2XS1i**,**u3XS1i**);**dXS1i**)
•**UXS1i**$^{-1}$•**T**[**UXS1i**]$^{-1}$•**T**[**C1i**]
+ ...
+ S[1,nm](**I3**[SECT(**xm**,**u1XSmi**,**u2XSmi**,**u3XSmi**)]
∗ux(**xm**|
SECT(**xm**,**u1XS1mi**,**u2XS1mi**,**u3XS1mi**);**dXS1m**)
•**UXS1i**$^{-1}$•**T**[**UXS1i**]$^{-1}$•**T**[**Cmi**]
+ S[1,n01](**c01i**
∗ux(**x01**)|SECT(**x01**,**u1XS01i**,**u2XS02i**,**u3XS03i**)
+ ...
+ S[1,n0m](**c0mi**
∗ux(**x0m**)|SECT(**x0m**,**u1XS0mi**,**u2XS0mi**,**u3XS0mi**)
+ S[1,n11]
(**D3**[SECT(**x11**,**u1XS11i**,**u2XS11i**,**u3XS11i**)]

∗(vx(**x11**|
SECT(**x11**,**u1XS11i**,**u2XS11i**,**u3XS11i**);**dXi1**)
•**C11j**]
\+ ...
\+ S[1,n1m]
(c1m∗**D3**[SECT(**x1m**,**u1XS1mi**,**u2XS1mi**,**u3XS1mi**)]
∗(vx(**x1**|
SECT(**x1m**,**u1XS1mi**,**u2XS1mi**,**u3XS1mi**);**dXi1**)
•**C1mj**]
\+ ...

where
f1 is continuous everywhere with continuous gradients;
and

Cij = lim (**D3**[**xi**]∗(**f**(**xi+dXS1i**,a1)|a1;**dXS1i**)
–**D3**[**xi**]∗(**f1**(**xi**,a1)|a1;**dXS1i**))
c0ij= lim **f**(**x0i+dXS1i**,a1)|
SECT(**x0i**,**u1XS1i**,**u2XS1i**,**u3XS1i**)
– **f1**(**x0i**(**r**,a1))
C1ij the impulse constant
are constants.

Dually,
f(**r**(**x**,a1))
= **f1**(**r**(**x**,a1))
\+ S[1,n1](**I3**[SECT(**r1**,**u1RS1i**,**u2RS1i**,**u3RS1i**)]
∗(ur(**r1**|
SECT(**r1**,**u1RS1i**,**u2RS1i**,**u3RS1i**);**dRS1i**)
•**URS1i**$^{-1}$•**T**[**URS1i**]$^{-1}$•**T**[**C1i**]
\+ ...
\+ S[1,nm](**I3**[SECT(**rm**,**u1RSmi**,**u2RSmi**,**u3RSmi**)]
∗ur(**rm**|
SECT(**rm**,**u1RS1mi**,**u2RS1mi**,**u3RS1mi**);**dRS1m**)
•**URS1i**$^{-1}$•**T**[**URS1i**]$^{-1}$•**T**[**Cmi**]
\+ S[1,n01](**c01i**
∗ur(**r01**)|
SECT(**r01**,**u1RS01i**,**u2RS02i**,**u3RS03i**)

+ ...
+ S[1,n0m](**c0mi**
∗ur(**r0m**)|SECT(**r0m**,**u1RS0mi**,**u2RS0mi**,**u3RS0mi**)
+ S[1,n11]
(**D3**[SECT(**r11**,**u1RS11i**,**u2RS11i**,**u3RS11i**)]
∗(vr(**r11**|
SECT(**r11**,**u1RS11i**,**u2RS11i**,**u3RS11i**);**dRi1**)
•C11j]
+ ...
+ S[1,n1m]
(c1m∗**D3**[SECT(**r1m**,**u1RS1mi**,**u2RS1mi**,**u3RS1mi**)]
∗(vr(**r1**|
SECT(**r1m**,**u1RS1mi**,**u2RS1mi**,**u3RS1mi**);**dRi1**)
•C1mj]
+ ...

where

f1 is continuous everywhere with continuous gradients;

and

Cij = lim (**D3[ri]∗(f(ri+dRS1i**,a1)|a1;**dRS1i**)
–**D3[ri]∗(f1**(**ri**,a1)|a1;**dRS1i**))
c0ij = lim **f(r0i+dRS1i**,a1)|
SECT(**r0i**,**u1RS1i**,**u2RS1i**,**u3RS1i**)
– **f1**(**r0i**(**r**,a1))
C1ij the impulse constant

are constants."

In effect, functions of considerable complexity can be so analyzed.

Returning now to integration.

Sectional Integration

First integration of the functions themselves.

Definition (invergences, incurls, and ingradients)

Given

f(**x**(**r**,a1)), a measurable function over V3(**x**);
MX, a measurable set of V3(**x**);
pXS1 = pXS1∗**uXS1**
= p1XS1∗**u1XS1**+p2XS1∗**u2XS1**+p3XS1∗**u3XS1**,
a partition vector;

for

S[bn(MX∩PXS1n)], the sum
over which MX∩PXS1n is non–empty

I3[MX]•(**f**(**x**(**r**,ai));**dXS1**)
≡ lim(S[MX∩PXS1n](**f**(**x**(PXS1n(i,j,k)))•**mx**(PXS1n(i,j,k))))
= lim(S[MX∩PXS1n](**f**(**x**(PXS1n(i,j,k)))•**pXS1**/n^3))
= lim(S[MX∩PXS1n](**f**(**x**(PXS1n(i,j,k)))•**dXS1**))

I3[MX]∧(**f**(**x**(**r**,a1));**dXS1**)
≡ lim(S[MX∩PXS1n](**f**(**x**(PXS1n(i,j,k)))∧**mx**(PXS1n(i,j,k))))
= lim(S[MX∩PXS1n](**f**(**x**(PXS1n(i,j,k)))∧**pXS1**/n^3))
= lim(S[MX∩PXS1n]**f**(**x**(PXS1n(i,j,k)))∧**dXS1**))

I3[MX]∗(**f**(**x**(**r**,a1));**dXS1**)
≡ lim(S[MX∩PXS1n]**f**(**x**(PXS1n(i,j,k)))∗**mx**(PXS1n(i,j,k))))
= lim(S[MX∩PXS1n](**f**(**x**(PXS1n(i,j,k)))∗**pXS1**/n^3))
= lim(S[MX∩PXS1n](**f**(**x**(PXS1n(i,j,k)))∗**dXS1**))
as n→∞.

Dually

Given

f(**r**(**x**,a1)), a measurable function over V3(**r**);
MR, a measurable set of V3(**r**);
pRS1 = pRS1∗**uRS1**
= p1∗**u1RS1**+p2∗**u2RS1**+p3∗**u3RS1**,
a partition vector;

for

S[bn(MR∩PRS1n)], the sum
over which MR∩PRS1n is non–empty

I3[MR]•(**f**(**r**(**x**,a1));**dRS1**)

≡ lim S[bn(MR∩PRS1n)]**f**(**r**(PRS1n(i,j,k)))

•**m1**(PRS1n(i,j,k)))

= lim S[bn(MR∩PRS1n)](**f**(**r**(PRS1n(i,j,k)))•**pRS1**/n^3)

= lim(S[MX∩PX1n](**f**(**x**(PX1n(i,j,k)))•**dX1**))

I3[MR]∧(**f**(**r**(**x**,a1));**dRS1**)

≡ lim S[bn(MR∩PRS1n)](**f**(**r**(PRS1n(i,j,k)))

∧**m1**(PRS1n(i,j,k)))

= lim S[bn(MR∩PRS1n)](**f**(**r**(PRS1n(i,j,k)))∧**pRS1**/n^3)

= lim(S[MX∩PX1n](**f**(**x**(PX1n(i,j,k)))∧**dX1**))

I3[MR]∗(**f**(**r**(**x**,a1));**dRS1**)

≡ lim S[bn(MR∩PRS1n)](**f**(**r**(PRS1n(i,j,k)))

∗**m1**(PRS1n(i,j,k)))

= lim S[bn(MR∩PRS1n)](**f**(**r**(PRS1n(i,j,k)))∗**pRS1**/n^3)

= lim(S[MX∩PX1n](**f**(**x**(PX1n(i,j,k)))∗**dX1**))

as n→∞.

end of definition

The following results are given without accompanying proofs which may be found in *tp1.4:*

When the local set is mutually related to the material set the relationships between their integrals can be expressed.

For sectional gradients.

Theorem (integration of sectional gradients over a measurable set)

Given at observation a1

MR, a measurable set in V3(**r**);

SMR the surface of MR;

pRS1 = p1RS1∗**u1RS1**+p2RS1∗**u2RS1**

+p3RS1∗**u3RS1,** a partition vector;

{PRS1n}, a set of partitions defined from **pRS1**;

SMRPE|**pRS1**, the positive exterior surface of MR;

SMRPI|**pRS1**, the positive interior surface of MR;

f(**r**(**x**,a1)) with a basic continuous

sectional gradient over MR;

then

I3[MR]•(**D3**[**r**]∗(**f**|a1;**dRS1**)•[**URS1**]$^{-1}$•**T**[**URS1**]$^{-1}$;**dRS1**)

= lim S[bn(SMRPE|**pRS1**∩PRS1n)](**f**(**r**(PRS1n(i,j,k))+**dRS1**))

– S[bn(SMRPI|**pRS1**∩PRS1n)](**f**(**r**(PRS1n(i,j,k))))

≡ **I**[SMRPI|**pRS1**,SMRPE|**pRS1**](**f**(**sr**)).

Dually

Given

MX, a measurable set in V3(**x**);

pXS1 = p1XS1∗**u1XS1**

+p2XS1∗**u2XS1**+p3XS1∗**u3x1**, a partition vector;

PXS1n, a set of partitions defined from **pXS1**;

SMXPE|**pXS1**, the positive exterior surface of MX;

SMXPI|**pXS1**, the positive interior surface of MX;

f(**x**(**r**,a1)) with a basic continuous

sectional gradient over MX;

then

I3[MX]•(**D3**[**x**]∗(**f**|a1;**dXS1**)•[**UXS1**]$^{-1}$•**T**[**UXS1**]$^{-1}$;**dXS1**)

= lim S[bn(SMXPE|**pXS1**∩PXS1n)](**f**(**r**(PXS1n(i,j,k))+**dXS1**))

– S[bn(SMXPI|**pXS1**∩PXS1n)](**f**(**r**(PXS1n(i,j,k))))

≡ **I**[SMXPI|**pXS1**,SMXPE|**pXS1**](**f**(**sx**)).

Corollary

I3[MR]•(**D3**[**r**]∗(**f**|a1;**dRS1**)•[**URS1**]$^{-1}$•**T**[**URS1**]$^{-1}$;**dRS1**)

= **I3**[MX]•(**D3**[**x**]∗(**f**|a1;**dXS1**)

•[**UXS1**]$^{-1}$•**T**[**UXS1**]$^{-1}$;**dXS1**)

I[SMRPI|**pRS1**,SMRPE|**pRS1**](**f**(**sr**))

= **I**[SMXPI|**pXS1**,SMXPE|**pXS1**](**f**(**sx**)).

In this context step functions can be used in many combinations. See Appendix.

In summary, for Theoretical Physics surfaces need to be defined from observables as a prerequisite for producing scientific results. So done, it offers a great variety of functions which may be useful for describing and understanding the observables. Theoretical Physics also provides a large variety of integrals over the surfaces. The result then become a kind of calculus of the sun.

Tasks

Define and Analyze numerous surfaces of sun

Determine what levels of continuity various functions of these surfaces possess.

Determine the integrated results of these functions over the defined surfaces.

The color of the sun

We observe the color of the sun directly, and indirectly with instruments.

Observed directly the sun appears to have a whitish color. When refracted through a prism however, a beam of this whitish sunlight is spread through a spectrum of many colors from red to violet, the colors of a rainbow. Within this spectrum may be seen darkened lines.

The sun's spectrum is seen as the following.

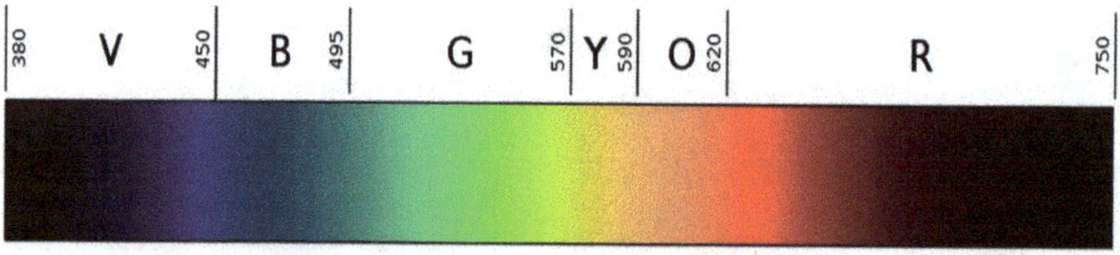

The colors are banded as violet (V), blue (B), green (G), yellow (Y), orange (O) and red (R) with marker wavelengths (in nanometers) from high to low frequencies.

The darkened lines are shown as

the spectrum of gaseous hydrogen. Close observation shows the lines (called the Balmer series) are measured as

656.3

486.1

434.0

410.2

397.0

where the numbers again represent wavelength in nanometers.

Spectrometers reveal radiation observable beyond the limits of visible light.

The results are shown here.

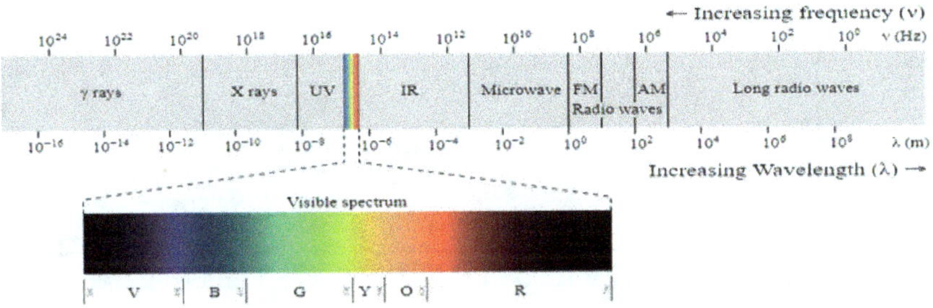

In this larger view, the solar radiation, as observed from earth is shown here

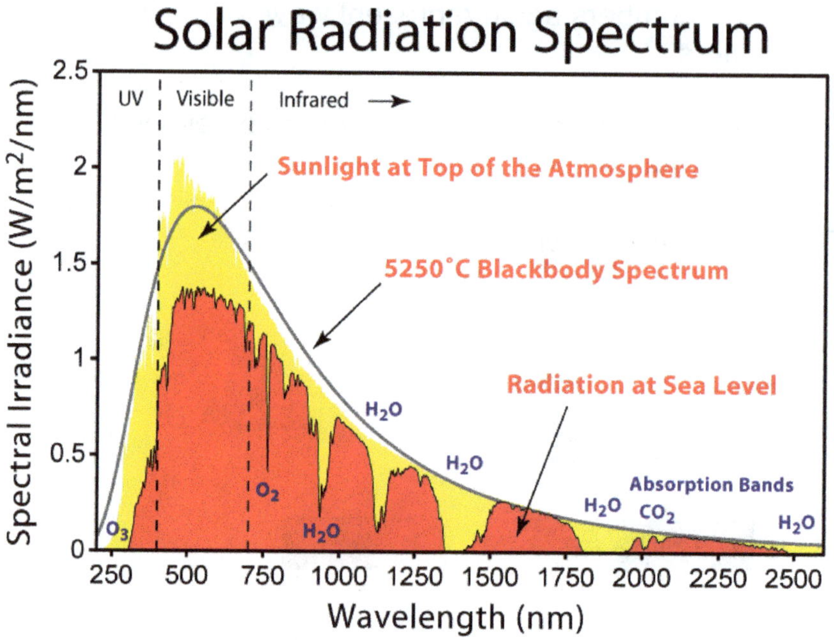

The spectrum received by satellites at the top of the atmosphere can be matched somewhat to a black body model at 5250 degrees centigrade, except for wavelengths less than 600 nanometers (higher frequencies). There a marked spikiness may be observed. The spectrum when observed at sea level differs markedly showing the effects of absorption by various components in the earth's atmosphere. The received spectrum at sea level can no longer be said to conform to the black body model.

Contemporary Explanations

Physicists have sought to explain these matters in terms of the radiation of a black body. So let us explain what is meant by a "black body."

Black bodies

The idea of a black body was introduced originally by Gustav Kirchhoff in 1860 as follows:

> "the supposition that bodies can be imagined which, for infinitely small thicknesses, completely absorb all incident rays, and neither reflect nor transmit any. I shall call such bodies *perfectly black*, or, more briefly, *black bodies*."

Kirchhoff's idea is clearly an imaginary ideal creation without relevance to physical bodies.

In 1900 the German physicist Max Planck heuristically derived a formula for the observed spectrum of heated bodies which he also referred to as the radiation from a black body. In effect Planck used Kirchhoff's appellation but changed its meaning. The result is confusion about the meaning of the words "black body."

The classical physics of Isaac Newton is incapable of explaining why some physical bodies begin to glow as they are heated.

Terrestrial experiments reveal that

- a sufficiently hot *gas* produces light with specific colors. These correspond to a spectrum with lines at discrete wavelengths.
- a sufficiently hot *solid* material produces light with a continuous spectrum.
- if the light from a hot glowing solid material passes through a gas of a cooler temperature then the spectrum has the discrete wavelengths characteristic of the gas deleted from the continuous spectrum of the material.

In solids the effect is easily observed, say in the manufacture of steel. At low temperatures no radiation is observable, but as the bolt of metal is heated to higher and higher temperatures it begins to glow first a deep red, then a greenish white turning to a brilliant incandescent white. The observation becomes even more spectacular when aided by a spectroscope.

The instrument then reveals that even at lower temperatures when no radiation is observable to the human eye, the bolt is still emitting radiation but at frequencies less than visible light. Importantly, a continuum of frequencies is observed whose spectrum has a definite shape.

As the temperature increases, the frequencies not only shift higher, but the shape of the spectrum also changes from broader and less intense to more peaked and more intense. Further complicating the matter, radiation from physical bodies varies somewhat with the body's chemical composition and the environment, all the while following the general trend described above.

Modern physicists proposed a mathematical model to describe these observations. They called the model a "black body". The appellation is unfortunate since the mathamatical "black body" is not a physical body at all, but just a mathematical equation which purports to describe the radiation of a fictitious body.

The idea of a mathematical black body:

1. excludes the effects of convection, conduction, and radiation from external sources;
2. assumes the fictitious body in thermal equilibrium with its environment;
3. assumes a chemically homogeneous body;
4. radiates according to the equation
   $\text{janskies} = (\text{constant1} * \text{frequency}^3)$
   $*(1/((\exp(\text{constant2} * \text{frequency}/\text{temperature}) - 1)))$

The black body equation is easily seen as composed of two factors:

a cubic function of the transmitted frequencies

(constant1∗frequency^3)

and a complicated shaping function

(1/((exp(constant2∗frequency/temperature) −1))).

The equation is expressed in terms of frequency and temperature.

The cubic function has the effect of moving the strength of the radiation increasingly higher with increasing frequency. The second function is an intricate shaping function to accommodate the changes in the shape of the radiated spectrum as both temperature and frequency vary.

The function is controlled by two constants. The first constant, constant1, is given as

constant1 = 2∗h/c^2

where h is Planck's constant (6.62606957×10^{-34} Jansky-seconds-meters^2) and *c* is the accepted constant speed of light (299,792,458 meters/second).

The second constant, constant2 is given as

constant2 = h/k

where h is again Planck's constant and k is Boltzmann's constant (1.3806488×10^{-23} cubic meters).

The combination h∗ν denotes energy in a 1 Hz band[1]. The combination k∗T where T denotes temperature in degrees Kelvin) denotes the average kinetic energy of one particle in an ideal gas. The equation is called Planck's Law and is variously written as

1. In conformity with the literature, the Greek letter ν is used to designate frequency; the symbol T is used to designate absolute temperature (degrees Kelvin). The mixture of Greek letters, caps and lower case Latin characters appears arbitrary.

$$B(\nu,T)=(2*h*\nu^3/c^2)*(1/(\exp(h*\nu/k*T)-1)$$

or

$$E_\lambda = \frac{8\pi hc}{\lambda^5} \times \frac{1}{\exp(hc/kT\lambda) - 1}.$$

where λ is wavelength.

The following table of physical units will help clarify the ideas and substitutions in the equations.

Variable	Physical Unit	Comment
J	F*V/(V*L) =F/L	Jansky
h	F*L*L/V=F*L*t	Planck's constant $6.62606957*10^{-34}$
c	V	299,792,458 m/s
t	L/V	time
ν	V/L = 1/t	frequency
l	L	Wave length
T	$F*L/L^3 = F/L^2$	temperature
k	$F*L/(F/L^2)=L^3$	Boltzmann's constant $1.3806488*10^{-23}$ cubic meters
h*ν	$F*L^2/V)*V/L = F*L$	energy
k*T	$L^3*F/L^2 =F*L$	energy
h/c^2	$F*L*L/V^3 = F*t^2/V$	constant
$h*\nu^3/c^2$	$F*L*(V/L)^2/V^2 = F/L$	a gradient of force
k/h	L*V/F	$0.208268263*10^{+11}$
h/k	F/(L*V)	$4.801499688*10^{-11}$
h/kT	L/V	time

Physical Units of Black Body Variables

A jansky has physical units of F/L = (F*V*t)/(L*L) (Watts*t/m^2) while Intensity has units of (F*V)/(L*L) = F/(L*t) (Watts/m^2).

Consequently in terms of physical units a Jansky is Intensity*time

It is easy to see that the formula (somewhat arrogantly called a Law) is merely a curve fitting exercise useful perhaps but providing little insight into understanding why the body radiates.

The following is a typical graph of the black body equation for various values of temperature.

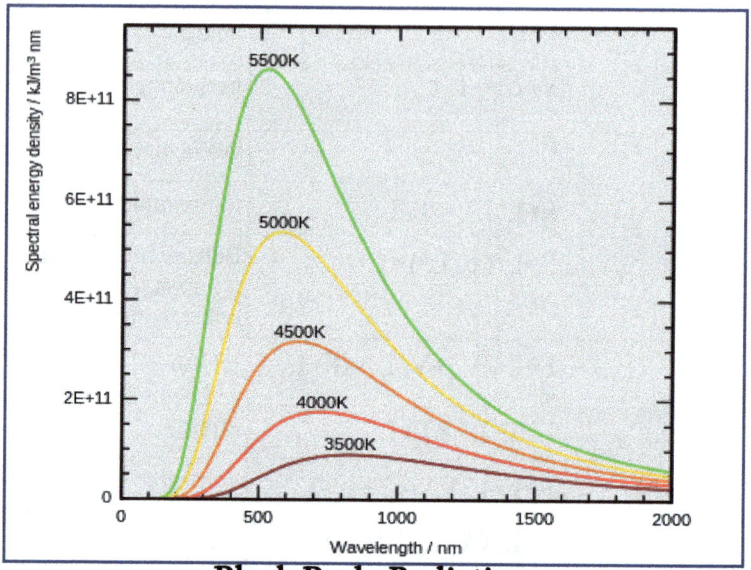

Black Body Radiation

The graph contains several assumptions. The X axis is given in wavelengths (λ) assuming the speed of light (c) constant using the following equation.

wavelength= c/frequency

The X axis is multiplied by 10^9 to obtain a result in nanometers.

Wavelengths are not directly measured, but only inferred. Frequency *is* directly measurable. In the graph frequencies increase unconventionally to the left.

The Y axis is given in scaled Janskies (radiated power per unit solid angle per unit frequency). A Jansky has a measured unit equal to 10^{-26} watts per meter^2 per Hz.

The black body model conforms only approximately to actual physical bodies. For terrestrial objects, both temperature and the radiated spectrum can usually be measured independently, and consequently the error of the black body model can then be calculated.

Can a scientific approach give an explanation of the observables and so possibly supplant the current models?

The scientific approach

Let us start by recognizing the central idea of temperature in this discussion.

Temperature

So what *is* temperature? We can measure it easily enough in most cases, but what are we measuring?

We need an idea of temperature that is related to the primary variables of theoretical physics: extension, motion, and force.

Heat is commonly and easily confused with temperature. Take a cup half filled with boiling water. We say: "The cup is hot." Now finish filling the cup with cold water. We say: "The cup is no longer hot." Where did the heat go?

To clarify our thinking, let us make a preliminary definition[1] of heat as a kind of energy. For the example above, in its first state (half full and boiling), the cup contained some heat. In the second state (full) the cup still contained all the heat it had previously. The "heat" did not go away; it was merely spread over a larger volume.

---

1. Temperature is something we can measure. Heat and energy are ideas of theoretical physics.

What changed then? This is where the idea of temperature comes in. In its first state the cup had a high temperature; in second state its temperature was lowered.

Temperature, then, has to do with energy in a *volume*. In terms of physical units[1],

units of temperature = F∗L/(L∗L∗L)

where F∗L are the physical units of energy and L∗L∗L are the physical units of volume.

Temperature and pressure are radically interrelated as may be seen by contemplating their physical units.

The physics units of pressure are

units of pressure = F/(L∗L)

which are just those of temperature.

So we can look upon temperature as a kind of pressure or pressure as a kind of temperature.

In Theoretical Physics energy is most generally described as a matrix with associated curls and divergences.

$$\mathbf{E} = \mathbf{f} * \mathbf{r}$$
$$\mathbf{e} = \mathbf{f} \wedge \mathbf{r}$$
$$e = \mathbf{f} \bullet \mathbf{r}$$

where **f** is some force, and **r** is some extension. Both the divergence and curl are included in **E** as its trace and off-diagonal elements respectively.

The elementary idea of force[2] comes in many varieties. The schema below is that contemplated by Theoretical Physics.

1. See *Theoretical Physics: the First Problem,*
2. The discussion of forces occupies a central position in Theoretical Physics. See *Theoretical Physics: the Second Problem*

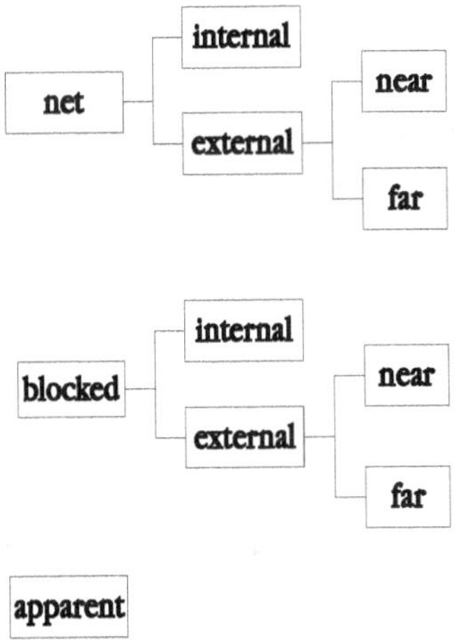

Schema of Forces

Net forces for theoretical physics are those which actually change movement. The same force, however, may effect different movement in different particles. The function $z(\mathbf{r}(\mathbf{x},a))$ proportions forces to movement. The following eight different net forces may then be defined:

$$\mathbf{fxxx} \equiv D[zx1(a)](\mathbf{vr}(\mathbf{x})|\mathbf{x1};dzx1)$$
$$\mathbf{fxxr} \equiv D[zx1(a)](\mathbf{vr}(\mathbf{x})|\mathbf{x1};dzr1)$$
$$\mathbf{fxrx} \equiv D[zx1(a)](\mathbf{vr}(\mathbf{x})|\mathbf{r1};dzx1)$$
$$\mathbf{fxrr} \equiv D[zx1(a)](\mathbf{vr}(\mathbf{x})|\mathbf{r1};dzr1)$$
$$\mathbf{frxx} \equiv D[zr1(a)](\mathbf{vx}(\mathbf{r})|\mathbf{x1};dzx1)$$
$$\mathbf{frxr} \equiv D[zr1(a)](\mathbf{vx}(\mathbf{r})|\mathbf{x1};dzr1)$$
$$\mathbf{frrx} \equiv D[zr1(a)](\mathbf{vx}(\mathbf{r})|\mathbf{r1};dzx1)$$
$$\mathbf{frrr} \equiv D[zr1(a)](\mathbf{vx}(\mathbf{r})|\mathbf{r1};dzr1)$$

These forces are designated "particle" or "point" according to their descriptors. For example **fxxx** is the particle-particle-particle force; **frrx** is the point -point-particle force, etc. The particle-particle-particle force, **fxxx**, is force on the particle observed causing the change in the observed particle velocity, **c**(**x1**,a), at **r1**(**x1**,a1). The observational mode implied in this sequence is material throughout. The point-point-point force, **frrr**, is force at the point causing the change in the point velocity, **vx** at **r1**(**x1**,a1). The observational mode implied in this sequence is local throughout. These forces, then, represent various modes of observing position, velocity, and forces.

Our discussion of temperature will then necessarily involve energy and volume, and then extend more profoundly into a discussion of forces.

In contemporary physics, temperature is associated only with scalar energy. In science, however, temperature is a matrix.

[**TEMP**] = **E**/volume

Temperature can change by changing the energy or the volume or both.

Using the calculus[1], the change in temperature can occur generally as

$$D[\mathbf{TEMP}] = ((D[\mathbf{E}]/\text{volume}) - \mathbf{E}*(D(\text{volume}))/(\text{volume})^2$$
that is
$$D[\mathbf{TEMP}]*(\text{volume})^2 = D[\mathbf{E}]*\text{volume} - \mathbf{E}*D(\text{volume})$$

In equilibrium when no change in temperature occurs,
$$D[\mathbf{E}] = \mathbf{E}*D(\text{volume})/(\text{volume})$$
that is
$$D[\mathbf{E}] = \mathbf{E}*D(\ln(\text{volume}))$$
where ln is the natural logarithm.

When the energy is merely scalar (trace(**E**)

---

1. The calculus provides the formula for f = u/v

$$df = (v*du - u*dv)/v^2$$

(change in energy)/volume

$$= \text{energy} * (\text{change in volume})/(\text{volume})^2$$

that is,

(change in energy)/energy = (change in volume)/volume.

The factor ((change in volume)/volume) is easily measured macroscopically; it is called the **coefficient of expansion**. A glance at the tables of expansion coefficients in a handbook of physical materials will quickly reveal the complexity of actual physical bodies.

It follows that whenever a physical body is held at a constant temperature macroscopically, it may nevertheless experience changes in its volume and its energy microscopically. The changes may occur in the body as a whole and in its separate parts.

The same argument holds for matrix energy **E**.

A satisfactory definition of temperature of a body has to specify the energy (of which there are many kinds) and the volume.

Let us take the volume first.

Volume can be thought of as macroscopic or microscopic. A large volume can be thought of as composed of many smaller volumes. There is little advantage to this decomposition unless the microscopic volume can be said to possess a property that the macroscopic volume does not.

When discussing properties of materials in terms of their coefficient of expansion, for instance, only the macroscopic volume is considered. The microscopic volumes escape interest. The variations of microscopic volumes are usually too difficult to measure.

For cases where the macroscopic volume is considered constant, however, the issue of microscopic volumes arises. Just because the macroscopic volume is constant, does it follow that the physical activity within the microscopic volumes is likewise constant? Is it not possible for the activity to vary from one microscopic volume to another such that their sum remains constant?

Thus a spectrum of microscopic volumes with different temperatures may be contemplated in physical materials.

Next, let's turn to the energy component of temperature.

Macroscopically, physical bodies are continually exchanging energy with their environments due to the microscopic motion at their boundaries.

Three types of motion are readily observable: ballistic, near, and remote. In ballistic motion the particle itself moves to a distant location. In near motion the particle moves other nearby particles. In remote motion the particle radiates its motion through a medium to influence a distant particle.

The near motion of purely mechanical particles give rise to mechanical convection and mechanical conduction; the remote motion of such particles is mechanical radiation (called sound).

Mechanical conduction and convection are physically similar. Conduction transfers mechanical motion from the body to an adjoining solid; convection transfers mechanical motion to an adjoining liquid, usually air.

Mechanical radiation is either coherent or not. Coherent radiation is usually considered as pressure[1] and treated as

---

1. The physical units of pressures are F/(L∗L) which are equivalent to those of temperature.

acoustic propagation. Incoherent radiation is considered as radiated temperature. Both required a medium to propagate.

The motion of purely charged particles is somewhat more complicated[1] because a moving charged particle gives rise to magnetic effects.

Similar to mechanical particles, charged particles may move ballisticly, by conduction, by convection, and by radiation. The sun's color is its electromagnetic radiation.

Theoretical Physics recognizes that radiation from a body occurs only from its surface. Radiation from a body is a surface[2] phenomenon as any painter will prove. Moreover observed radiation may not be a function of temperature at all.

How does temperature of a body relate to radiation from that body?

A scientific explanation must explain

- the radiation of a cold gas;
- the line radiation of a hot gas
- how heat in a solid results in electromagnetic radiation
- the radiation from solids composed of charged or non-charged particles.

in terms of the elementary ideas of physics: extension, motion, and forces.

---

1. See the discussion on Maxwell's equations in *Theoretical Physics: the Second Problem*
2. Physical surfaces are complicated affairs. They are discussed in *Theoretical Physics: the First Problem* and *tp1.3*. Surfaces may be closed, open, or anything in between.

Radiation of a cold gas

We do not see the air we breathe.

We see through the air we breathe and so observe radiation from light sources and their reflections. However, even in the absence of external radiation, the cold air will reveal a continuous spectrum when analyzed with a spectrometer. The spectrum contains absorption lines and as such is called a line absorption spectrum.

The radiation is often faint and lies outside the visible spectrum.

The input optics of the spectrometer provides a surface for measuring the radiation.

Radiation of a hot gas

Plasma is a state of matter which can be generated by heating a gas or subjecting it to a strong electromagnetic force. When the gas becomes a plasma, the spectrum ceases to be continuous and instead becomes a series of discrete lines. The lines correspond to the chemical composition of the gas, and they match the absorption lines in the spectrum of the cold gas.

The role of temperature

Consider a liquid in a bottle. Suppose the temperature of the liquid is due to the kinetic energy of the various particles in proportion to their volumes. The temperature of an individual particle **x** would then be

$$\mathrm{temp}(\mathbf{x}) = \mathrm{rden}(\mathbf{x}) * \mathbf{vr}(\mathbf{x}) \cdot \mathbf{vr}(\mathbf{x})/2$$

where rden(**x**) is the relative[1] density of **x**.and **vr**(**x**) is its material velocity. In the interior of the liquid the various

1. See *Theoretical Physics: the Second Problem* for definition and significance of this important idea.

particles collide with each other producing a spectrum of results. The relative size and relative density of the particles may be altered, as well as the associated forces, not only in magnitude and direction, but in type as well. As illustration, for certain conditions

$$\begin{aligned}\mathrm{rsz}(\mathbf{x1},\mathrm{a1}) = {} & \mathrm{D[a1]}(\mathbf{r}|\mathbf{x1};\mathrm{da})\bullet\mathbf{D3[r1]}*(\mathrm{rsz},\mathrm{a1}+\mathrm{da};\ \mathbf{dr})\\ & /(\mathrm{D3}[\mathbf{r1}]\bullet(\mathrm{D[a1]}(\mathbf{r}|\mathbf{x1};\mathrm{da})|\mathrm{a1};\ \mathbf{dr})\\ & + \mathrm{D3}[\mathbf{x1}]\bullet(\mathrm{D[a1]}(\mathbf{x}|\mathbf{r1};\mathrm{da})|\mathrm{a1};\ \mathbf{dx}))\end{aligned}$$

$$\begin{aligned}\mathrm{rdn}(\mathbf{r1},\mathrm{a1}) = {} & -\mathrm{D[a1]}(\mathbf{r}|\mathbf{x1};\mathrm{da})\bullet\mathbf{D3[r1]}*(\mathrm{rdn}|\mathrm{a1}+\mathrm{da};\ \mathbf{dr})\\ & /(\mathrm{D3}[\mathbf{r1}]\bullet(\mathrm{D[a1]}(\mathbf{r}|\mathbf{x1};\mathrm{da})|\mathrm{a1};\ \mathbf{dr})\\ & + \mathrm{D3}[\mathbf{x1}]\bullet(\mathrm{D[a1]}(\mathbf{x}|\mathbf{r1};\mathrm{da})|\mathrm{a1};\ \mathbf{dx}))\end{aligned}$$

These equations are an illustration of a fundamental theorem in Theoretical Physics called the expansion theorem[1].

The expansion theorem applies to any type of relative density, not as an observed discovery but as a logical consequence of theoretical physics.

At the surface of the bottle, however, something different happens. As the particle collides with the surface it is repelled by a force which counters the particle's force. We call this countering force a blocking force because it blocks the particle's motion from being transmitted beyond the bottle.

In contrast to the particle velocity which is located here and there at various times within the bottle, the blocking force, even as it dynamically reacts, has a fixed location at the container's surface.

We can infer the existence of blocking forces for all surfaces which maintain their integrity.

---

1. See *Theoretical Physics: the Second Problem* for proofs.

What happens as the temperature rises? Then **vr**(**x**) increases in magnitude. The force driving the particle may increase beyond the ability of the blocking force to counteract, and so something of **vr**(**x**) may escape the container.

So let us apply this understanding to mechanical radiation. As a first event suppose a particle interior to the container collides with a thin surface with sufficient force to deform the surface at the point of collision. A second event follows with a collision with a particle of a lower energy insufficient to overcome the blocking force. The surface will then return to its former state. The exterior surface of the container will have communicated to the surrounding mechanical medium an impulsive motion. As the process repeats itself, all the particles with energy surpassing the blocking force will transmit their motion to the medium while those with energies less than the blocking force will not.

The net force applied to the container wall at **r1** by the collision with **x1** with velocity **vr**(**x1**,t1) is

fxxx(**x1**,t1) = resx(**x1**)∗D[t1](**vr**(**x1**|**r1**;dt))

where resx(**x1**) is the restraint of the particle **x1**.

From a different mode of observation[1]

frrr(**r1**,t1) = resr(**r1**)∗D[(t)](**vx**(**r**)|**r1**;dt)
= **fxxx**(**x1**,t1)

The successive forces applied at **r1** by successive particles is described as

frrr(**r1**,t) = resr(**r1**)∗D[(t)](**vx**(**r**)|**r1**;dt)

which result from a succession of net forces

fxxx(**x**,t) = resx(**x**)∗D[t1](**vr**(**x**|**r1**;dt))

for particles **x**(**r1**,t).

---

1. A better explanation would include collisions. See *Theoretical Physics: the Second Problem.*.

Let **fb(x|r1**,t) be the dynamic blocking force.

Then for simplicity neglecting the transmission through the wall of the container, the force transmitted to the exterior of the container is

frxxr(**x**|**r1**,t) = resx(**x**)∗D[(t)](**vr**(**x**)|**r1**;dt) – **fb(x|r1**,t)

The movement at the exterior of the container launches a mirrored movement in the medium which is the propagation.

A spectrometer intercepting the propagation will sort them into a spectrum.

The resulting spectrum reflects the temperature of the particles within the body capable of exceeding the blocking force.

If the macroscopic temperature of the body is increased and stabilized, the spectrum changes to reflect the increased number of particles able to exceed the blocking force as well as the relative number of higher temperature impulses.

The situation is somewhat changed for charged particles. Again the mechanical motion of the charged particles will cause effects similar to effects caused by non-charged particles. In addition motion of the charge particle produces additional magnetic forces in the particles. These forces are also related to a temperature.

The energy described as a full matrix contains off diagonal elements which can be related to charged particles[1]. The external medium must then be able to transmit electromagnetic radiation.

---

1. See *Theoretical Physics: the Second Problem* and *Theoretical Physics: the Third Problem* for an explanation of Maxwell's equations in these terms.

The container may or may not react with additional blocking forces. Depending on the medium the propagation may be electromagnetic.

The complexity of possible combinations is shown in the following table

item	mechanical	electrical	both
8 sub types of net forces	☑	☑	☑
temperature	☑	☑	☑
pressure	☑	☑	☑
container	☑	☑	☑
blocking	☑	☑	☑
Spectrum mirroring particle motion.	☑	☑	☑

To summarize then, the radiation requires

- a surface with blocking forces
- interior particles moving with a spectrum of temperatures.
- an external medium

The particles need not be of the same size.

The radiation will then have a spectrum which shifts with temperature both in magnitude and frequency

It should be noted that the scientific explanation has no need for Planck's constant, Boltzmann's constant, or of the idea of photons all of which are seen merely as factors in engineering models. These factors imply homogenous microscopic volumes. It is to noted also that radiation is a function of conditions of a surface, not of bodies as a whole.

Line spectra

The lines in these spectra always occur in company with others and have about the same brightness.

When the particles in a container are raised to a sufficiently high temperature or subjected to sufficient external electromagnetic forces they become so ionized that positive and negative charges become disassociated.

Such a state is called a plasma. One may then talk of the microscopic temperature of the positive ions and the negative ions.

Plasmas induced by electromagnetic forces may exist at much lesser temperatures than those induced by temperature alone.

Since the spectrum consists of many discrete lines, the charged particles must have been many, varied and discrete also.

It should be noted that even in the case of a single chemical element, measurements always involve multiple particles.

In plasmas, it appears, internal blocked forces come into play. The process of plasmatizing the set of particles of a single element appears to unblock some of the blocked forces otherwise inherent in the element. The many lines of the spectrum of a single element indicates that the blocked forces of a single element are many and discrete[1]. The **frrr** force due to a single blocked force of identical ions acting on the container will then be identical in magnitude and vary only in direction. The corresponding force **frxxr**(**x**|**r1**,t) will

1. It does not follow, however, that each individual particle of the element contains all the blocked forces.

then propagate into an acceptable medium as a single frequency of constant amplitude.

The number, strength, and frequency of the observed lines in the element's spectrum, thus reveal the complexity of the blocked forces in the chemical element itself.

Multiple released blocked forces of the same particle may add and subtract vectorially. The result may produce other spectral lines.

The observed line spectra reveal the complexity of blocked forces of even a homogeneous grouping of a single chemical element[1].

Application to the sun

The reigning speculation about the sun claims that the sun is a plasma composed mostly of hydrogen. However the sun radiates a continuous spectrum, not lines.

The contradiction forces a radical change in our thinking about the sun.

Similarly plasmas are held to be by far the most common phase of ordinary matter in the universe, both by mass and by volume. Above the Earth's surface, the ionosphere is a plasma, and the magnetosphere contains plasma. Within our solar system, interplanetary space filled with the plasma expelled via the solar wind.

---

1. This is matter for a much more profound investigation. The reader may be well advised to start with the detailed investigation of an ideal physical wave contained in *Theoretical Physics: the Third Problem.*

Big Bang enthusiasts claim[1] cosmic microwave background (CMB) as evidence of an initial cosmic event. The observed spectrum is continuous, a good fit to a black body curve at 2.72548±0.00057 degrees Kelvin.

Again we see that a radiation from a supposed plasma is continuous, not discrete lines.

A logical problem exists here.

1. If it is true that

   radiation from a plasma implies discrete line spectra
   then the lack of a discrete line spectra does not
   necessarily imply
   that the radiation is not from a plasma.

2. If it is true that

   radiation from a plasma occurs *if and only if*
   the radiation is one of discrete line spectra
   then the lack of a discrete line spectra does imply
   that the radiation is not from a plasma.

Which statement is true?

How might Theoretical Physics suggest a resolution?

The analysis of an ideal physical wave in *Theoretical Physics: the Third Problem* shows that the density of the particles involved in propagation varies. (Changes in density involve volume and so temperature.)

Furthermore such propagation implies resonance and a stiffness in the medium.

Are these ideas relevant to the sun's radiation?

Theoretical Physics also notes that the relation of temperature to radiation is not necessarily fixed.

---

1. A claim easily refuted in *Playing with the Big Bang* ISBN 978-0-9844299-6-7.

Two quick instances make the point. A painter may change the spectrum of an object by painting it a different color without changing its temperature. The temperature of an incandescent bulb differs from the temperature of an LED bulb even when they emit the same spectrum.

Theoretical Physics provides a well-stocked armory of intellectually coherent ideas and statements with which to describe and understand observations. The present thinking about the sun appears to be an accumulation of engineering models running into a maze of contradictions unable to explain current observations. They provide no understanding of the observations.

It will be a great adventure for those physicists who choose to apply the ideas of Theoretical Physics rigorously to come to better understand their observations of the sun.

Tasks

1. The earth is located close to the plane of the sun's equator. Does the sun's spectrum differ when observed along the sun's axis?
2. Again, the sun with its atmosphere rotates. Does the rotation of the turbulent atmosphere cause the received spectrum of the sun to change?
3. Is the sun's radiation incoherent?
4. Do Maxwell's equations hold for the sun?. In particular, does the D**D3** interchange which underlies the Faraday equations hold for the sun?
5. Ditto for particle emissions from the sun.
6. Is the black body estimate of the sun's surface temperature correct?
7. Find a way to measure the temperature of the sun's surface independent of black body estimations.
8. Is the temperature of the sun's surface constant, or does it vary from place to place and with time?

9. If the surface temperature is unknown, are all conjectures about the interior of the sun suspect?
10. Calculate the temperature of the sun's surface from place to place and time to time.
11. What is the relationship between pressure and temperature on the sun's surface?
12. From observed movements of the particles of the sun's surface, infer net forces.
13. Describe the ballistic emissions of particles from the sun.
14. Describe the near motions of particles on the sun's surface.
15. What are the blocking forces at the surface of the sun which give a surface its shape?
16. Of the many different kinds of energy which ones are represented on the sun's surface?
17. What is the relative density of particles on the sun's surfaces?
18. Does the expansion theorem hold for the sun's surface?
19. Investigate radiation from plasmas of very low density and very high density.
20. Investigate the effect of resonance in a radiating source.
21. Investigate the effect of stiffness in a radiating medium.

Summary

When a theoretical physicist looks at the sun, he is first struck by the fact the most contemporary research on the sun is technology, not science. As such despite whatever utilitarian goals such research has, it does not produce understanding.

Is a science of the sun is possible? A fully developed science of the sun would include an explicit set of axioms, coherent definitions, an algebra, and a calculus. We are far from that at present.

But we do have the intellectual tools of Theoretical Physics with which to start.

When a theoretical physicist applies these tools to the sun, he finds it relatively easy to come to some understanding of what is observed. Yet easier still are the many questions that arise.

He is also encouraged by the increased resolution of instruments used to observe the sun. As fundamental inputs to a pursuit of understanding, these should greatly help the development of a science of the sun

This theoretical physicist also acknowledges that the investigating and answering of these many questions, and others yet unasked, lie beyond his humble capabilities. He therefore invites others to participate in the adventure of creating a science of the sun.

Appendix

A first task for the solar physicist will be to examine which regions of the sun support gradients of specific functions and which do not[1].

This appendix gives a synopsis of indexed derivatives followed by a synopsis of the gradients contemplated by Theoretical Physics and relationships between them. The reader will find a much more ample discussion in the foundational literature.

Derivatives in **V3** belong to functions rather than to relationships and do not necessarily arise from functional values. A smaller class of functions called **basic** possess derivatives from which an accurate estimate of neighboring functional values may be made.

The derivatives of Theoretical Physics comprehend

indexed derivatives
directional derivatives
directional gradients
sectional gradients

and may vary not only from location to location, particle to particle but also from observation to observation.

Please note the derivatives of a generalized function in Theoretical Physics need not exist, need not be basic, need not be continuous.

Indexed derivatives

A convention is needed to designate the situation before during and after the observation designated as *a1.*

1.. Given a specific continuous function, the set of look-alike functions which support a derivative in the Newtonian sense has measure zero. The derivatives of the primary variable **r**, **x** always exist as basic derivatives, although they may not be continuous. Cf. tp1.3:

The panoramic convention is shown in the following compounded table:

Index	s0	r0	r1	r2	s2		y0	x0	x1	x2	y2	Index
a1−da		x1	x0					r1	r0			a1−da
a1	x0	y0	x1	y2	x2		r0	s0	r1	s2	r2	a1
a1+da			x2	x1					r2	r1		a1+da

Panoramic Reference Designations

The convention shows the particle **x1** moving from **r0** to **r1** to **r2** while location **r1** is occupied by **x0** then **x1** then **x2**. Other particles and locations then come into view.

Differences may then be constructed from which derivatives may be defined as limits.

While arising from different definitions, all these differences can be reduced to four.

In possible contrast to other functions, in Theoretical Physics the derivatives of *primary* functions always exist as basic derivatives. They need not, however, be continuous.

The condition

$$D[a](\mathbf{r}|\mathbf{x1};da) = \mathbf{0}$$

is called **materially stalled**.

The joint conditions

$$D[a1](\mathbf{r}|\mathbf{x1};d_f a) = \mathbf{0},$$
$$D[a1](\mathbf{r}|\mathbf{x1};d_b a) \neq \mathbf{0}$$

are called **materially stopped** at observation *a1*.

The joint conditions

$$D[a1](\mathbf{r}|\mathbf{x1};d_f a) \neq \mathbf{0},$$
$$D[a1](\mathbf{r}|\mathbf{x1};d_b a) = \mathbf{0}$$

are called **materially started** at observation a1.

These derivatives, indexed on observations are called indexed derivatives.

Directional Gradients

Given suitable continuity, the four distinct, restricted increments of Theoretical Physics each approaches **0** as a1+da ⟶ a1 or a1−da ⟶ a1. Combined with the twelve index derivatives of Theoretical Physics, they can be used to define *forty–eight* distinct local/material directional gradients in **V3** in the form of:

$$\mathbf{D1}[\mathbf{r1}(\mathbf{x1},a1),\mathbf{r2}(\mathbf{x1},a1+da)]*(\mathbf{f}\,|\,\mathbf{x1};\mathbf{d}_b\mathbf{r})$$
$$\equiv \mathbf{D1}[\mathbf{r1},\mathbf{r2}]*(\mathbf{f}\,|\,\mathbf{x1};\mathbf{d}_b\mathbf{r})$$
$$= \lim\,[\mathbf{diff}_f(\mathbf{f}\,|\,\mathbf{x1})*\mathbf{qd}(\mathbf{d}_b(\mathbf{r}\,|\,\mathbf{x1}))].$$

where qd is the directional reciprocal vector.

Where a mixture of senses is contemplated, it is assumed

$$a1 + da - a1 = a1 - (a1 - da).$$

Directional gradients are more adequately described in *tp1.4* and further still in *Theoretical Physics: The First Problem*.

Theoretical Physics defines numerous gradients with possible applicability to the sun.

Let us focus first on gradients that may be observed during a single observation, leaving for later gradients that require more than one observation.

Directional gradients in one observation

The definition of **D1**∗(**f**|a;**dr**) calls for the selection of a direction **utr**. For the primary functions, **r** and **x**, there are 4∗4=16 such distinct directional gradients which Theoretical Physics contemplates.

Analysis can proceed for any given observation of the sun's surface.

- How far does the gradient continue constant for a given direction?
- Which areas of the sun's surface show consistent results for any chosen direction?
- Which areas of the sun's surface show similar patterns in all sixteen sectional gradients for a specific chosen direction?

Directions may also be taken into the surface and out of the surface. Sharp discontinuities in positive and negative directions may help define a surface of the sun distinctly.

Analysis of the sun by directional gradients promises many interesting results as may be surmised by applying a similar analysis to the earth.

Sectional Gradients

Sectional gradients may also be calculated from a single observation.

The definition of the sectional gradients for the primary variables is:

Definiton
  Given
    XS1 = SECT(**x1**,**u1XS1**,**u2XS1**,**u3XS1**)
    RS1 = SECT(**r1**,**u1RS1**,**u2RS1**,**u3RS1**)
      as linearized mutually related sections
        over which **r**(**x**,a1) and **x**(**r**,a1) are
          mutually differentiable at **r1**(**x1**,a1)
  then
D3[x1]∗(**r**(**x**,a)|a1;**dXS1**)
    ≡ lim (**r**(**x1**+d1XS1∗**u1XS1**) – **r**(**x1**))
          ∗**u1XS1**/d1XS1
      + (**r**(**x1**+d2XS1∗**u2XS1**) – **r**(**x1**))

∗**u2XS1**/d2XS1
\+ (**r**(**x1**+d3XS1∗**u3XS1**) – **r**(**x1**))
∗**u3XS1**/d3XS1

D3[**r1**]∗(**x**(**r**,a)|a1;**dRS1**)
≡ lim (**x**(**r1**+d1RS1∗**u1RS1**) – **x**(**r1**))
∗**u1RS1**/d1RS1
\+ (**x**(**r1**+d2RS1∗**u2RS1**) – **x**(**r1**))
∗**u2RS1**/d2RS1
\+ (**x**(**r1**+d3RS1∗**u3RS1**) – **x**(**r1**))
∗**u3RS1**/d3RS1
end of definition

In Theoretical Physics the basic gradients of primary functions always exist. By the principle of non–collocation the gradients of primary functions always have rank 3.

The complexity is shown in the following table.

<table>
<tr><th>Transformation</th><th>By Section</th><th>By Gradient</th></tr>
<tr><td>Local Octant
to Local Octant</td><td>URj,Rk</td><td>dRj • [URj,Rk]$^{-1}$
•T[D3[r1]∗(r(x,a)
|a1;dr$_k$]</td></tr>
<tr><td>Material Octant
to Material Octant</td><td>UXj,Xk</td><td>dXj • [UXj,Xk]$^{-1}$
•T[D3[x1]∗(x(r,a)
|a1;dx$_k$]</td></tr>
<tr><td>Local Octant
to Material Octant</td><td>URj,Xk</td><td>dRj •T[D3[r1]∗(x(r,a)
|a1;drj)]</td></tr>
<tr><td>Local Octant
to Local Section</td><td>URj,RSk</td><td>dRj •T[URj,RS]$^{-1}$
•T[D3[r1]∗(r(x,a)
|a1;dRS)]</td></tr>
<tr><td>Local Octant
to Material Section</td><td>URj,XSk</td><td>dRj •T[D3[r1]∗(x(r,a)
|a1;dr$_j$)]</td></tr>
<tr><td>Material Octant
to Material Section</td><td>UXj,XSk</td><td>dXj •T[D3[x1]∗(r(x,a)
|a1;dx$_j$)]</td></tr>
<tr><td>Material Octant
to Local Section</td><td>UXj,RSk</td><td>dXj •T[D3[x1]∗(r(x,a)
|a1;dx$_j$)]</td></tr>
<tr><td>Local Section
to Local Section</td><td>URSj,RSk</td><td>dRSj • URSj]$^{-1}$]
•T[URSk^{-1}]</td></tr>
</table>

Transformation	By Section	By Gradient
		•T[[D3[**r1**]∗(**r**(**x**,a) |a1;**dRSk**)]
Local Section to Material Section	URSj,XSk	**dXS= dRS** •[**URS**$^{-1}$]•T[**URS**$^{-1}$] •T[D3[**r1**]∗(**x**(**r**,a) |a1;**dRS**)]
Material Section to Material Section	UXSj,Xsk	**dXSj** • **UXSj**]$^{-1}$] •T[**UXSk**$^{-1}$] •T[[D3[**x1**]∗(**x**(**r**,a) |a1;**dXSk**)]

The conditions underlying the results are:

1. the gradient must be basic
2. The symbology uses the positive/definite convention
3. material and local sections must be related linearly
4. the local and material sections under **r**(**x**.a) are mutually dependent
5. **r**(**x**.a) and **x**(**r**.a)must also be mutually differentiable

When more than one observation is available, results pertaining to the change in observations are available.

Relationships between the Derivatives of Theoretical Physics

The derivatives of Theoretical Physics arise from functions of the one given set of observations, **r**(**x**,a), alternately stated as **x**(**r**,a). Consequently one might expect related observations to generate relationships among derivatives.

Results with implied reference of **r1**(**x1**,a1) are summarized in the following tables.

Derivatives	Rank 1 Decomposition
D[r0,r1](f|a1;da)	D(r|x1;d_ba) • T[D1[r0(x,a1),r1(x1,a1)]∗(f|a1;d_br)]
D[r0,r1] (f|a1−da;da)	D(r|x1;d_ba) • T[D1[r0(x1),r1(x,a1−da)]∗(f;d_br)]
D[r1,r2](f|a1+da;da)	D(r|x1;d_fa) • T[D1[r1(x,a1+da),r2(x1)]∗(f;d_fr)]
D[(r1,r2](f|a1;da)	D(r|x1;d_fa) • T[D1[r1(x1),r2(x,a1)]∗(f|a1;d_fr)]
D[x0,x1](f|a1;da)	D(x|r1;d_ba) • T[D1[x0(r,a1),x1(r1)]∗(f;d_bx)]
D[x0,x1](f|a1−da;da)	D(x|r1;d_ba) • T[D1(x0(r1),x1(r1,a1−da)]∗(f;d_bx)]
D[x1,x2](f|a1+da;da)	D(x|r1;d_fa) • T[D1(x1(r,a1+da),x2(r1)]∗(f|a1+da;d_fx)]
D[x1,x2](f|a1;da)	D(x|r1;d_fa) • T[D1[x1(r1),x2(r,a1)]∗(f|a1;d_fx)]

Decomposition of Indexed Derivatives

Derivatives	Rank 3 Decomposition
D[r0,r1](f|a1;da)	D(r|x1;d_ba) • T[URSb(a1)$^{-1}$] • [URSb(a1)^{1}] • T[D3[r1]∗(f|a1;dRSb)]
D[r0,r1](f|a1−da;da)	D(r|x1;d_ba) • T[URSb(a1−da)$^{-1}$] • [URSb(a1−da)$^{-1}$] • T[D3[r1]∗(f|a1−da;dRSb)]
D[r1,r2](f|a1+da;da)	D(r|x1;d_fa) • T[URSf(a1+da)$^{-1}$] • [URSf(a1+da)$^{-1}$] • T[D3[r1]∗(f|a1+da;dRSf)]
D[r1,r2](f|a1;da)	D(r|x1;d_fa) • T[URSf(a1)$^{-1}$] • [URSf(a1)$^{-1}$] • T[D3[r1]∗(f|a1;dRSf)]
D[x0,x1](f|a1;da)	D(x|r1;d_ba) • T[UXSb(a1)$^{-1}$] • [UXSb(a1)$^{-1}$] • T[D3[x1]∗(f|a1;dXSb)
D[x0,x1](f|a1−da;da)	D(x|r1;d_ba) • T[UXSb(a1−da)$^{-1}$] • [UXSb(a1−da)$^{-1}$] • T[D3[x1]∗(f|a1−da;dXSb)]
D[x1,x2](f|a1+da;da)	D(x|r1;d_fa) • T[UXSf(a1+da)$^{-1}$] • [UXSf(a1+da)$^{-1}$] • T[D3[x1]∗(f|a1+da;dXSf)]
D[x1,x2](f|a1;da)	D(x|r1;d_fa) • T[UXSf(a1)$^{-1}$] • [UXSf(a1)$^{-1}$] • T[D3[x1]∗(f|a1;dXSf)]

Decomposition of Indexed Derivatives

The Bridge Theorem

The relationship between material and local references is clarified in the following important theorem.

Theorem (bridge theorem: first form)

Given

three ordered observations designated

r(**x**,a1–da), **r**(**x**,a1) and **r**(**x**,a1+da)

corresponding to

x(**r**,a1–da) **x**(**r**,a1) and **x**(**r**,a1+da);

three functions of these observations having similar physical units

f(**r**(**x**,a1–da)), **f**(**r**(**x**,a1)), and **f**(**r**(**x**,a1+da));

for

r1(**x1**,a1) the location of the reference particle **x1**(**r1**,a1);

r0 ≡ **r**(**x1**,a1–da);

r2 ≡ **r**(**x1**,a1+da);

x0 ≡ **x**(**r1**,a1–da);

x2 ≡ **x**(**r1**,a1+da);

s0 ≡ **r**(**x0**,a1);

s2 ≡ **r**(**x2**,a1);

y0 ≡ **x**(**r0**,a1);

y2 ≡ **x**(**r2**,a1);

then

f(**r**(**x1**,a1+da))

= **f**(**r1**(**x**,a1+da)) + **f**(**r2**(**x**,a1+da)) – **f**(**r1**(**x**,a1+da))

f(**r**(**x1**,a1))

= **f**(**r2**(**x**,a1)) + **f**(**r1**(**x**,a1)) – **f**(**r2**(**x**,a1))

f(**x**(**r1**,a1+da))

= **f**(**x1**(**r**,a1+da)) + **f**(**x2**(**r**,a1+da)) – **f**(**x1**(**r**,a1+da))

f(**x**(**r1**,a1))

= **f**(**x2**(**r**,a1))+ **f**(**x1**(**r**,a1)) – **f**(**x2**(**r**,a1))

and

f(r(x1,a1))
= f(r0(x,a1)) + f(r1(x,a1)) – f(r0(x,a1))
f(r(x1,a1–da))
= f(r1(x,a1–da)) + f(r0(x,a1–da)) – f(r1(x,a1–da))
f(x(r1,a1))
= f(x0(r,a1)) + f(x1(r,a1)) – f(x0(r,a1))
f(x(r1,a1–da))
= f(x1(r,a1–da)) + f(x0(r,a1–da)) – f(x1(r,a1–da)).

Except for compatibility of units, no restrictions are placed on the functions, **f**(a1–da), **f**(a1), and **f**(a1+da). The theorem holds even for stalled processes.

Now consider the three functions above as three instances of a single function, **f**(**r**(**x**,a))=**f**(**x**(**r**,a)) over the real index a. This consideration leads to a second form of the bridge theorem.

Theorem (bridge theorem: second form)
Given
r1(**x1**,a1) the location of the reference particle **x1**(**r1**,a1);
f(**r**(**x**,a));
for
diff$_n$(f(r(x,a))|q) the change in **f** from **f**(**r1**(**x1**,a1))
with q as reference
in either a backward or forward sense
with n symbolizing forward or backward
then
diff$_f$(f|x1) = diff$_f$(f|r1) + diff$_f$(f(r|a1+da))
= diff$_f$(f|r2) + diff$_f$(f(r|a1))
diff$_f$(f|r1) = diff$_f$(f|x1) + diff$_f$(f(x|a1+da))
= diff$_f$(f|x2) + diff$_f$(f(x|a1))
diff$_b$(f|x1) = diff$_b$(f|r1) + diff$_b$(f(r|a1–da))
= diff$_b$(f|r0) + diff$_b$(f(r|a1))
diff$_b$(f|r1) = diff$_b$(f|x1) + diff$_b$(f(x|a1–da))
= diff$_b$(f|x0) + diff$_b$(f(x|a1))

which imply

$$\mathbf{diff_f}(\mathbf{f}(\mathbf{r}|a1+da)) = -\mathbf{diff_f}(\mathbf{f}(\mathbf{x}|a1+da))$$
$$\mathbf{diff_b}(\mathbf{f}(\mathbf{r}|a1-da)) = -\mathbf{diff_b}(\mathbf{f}(\mathbf{x}|a1-da))$$
$$\mathbf{diff_f}(\mathbf{f}|\mathbf{x1}) - \mathbf{diff_f}(\mathbf{f}|\mathbf{x2}) = \mathbf{diff_f}(\mathbf{f}(\mathbf{r}|a1+da)) + \mathbf{diff_f}(\mathbf{f}(\mathbf{x}|a1))$$
$$\mathbf{diff_b}(\mathbf{f}|\mathbf{x1}) - \mathbf{diff_b}(\mathbf{f}|\mathbf{x0}) = \mathbf{diff_b}(\mathbf{f}(\mathbf{r}|a1-da)) + \mathbf{diff_b}(\mathbf{f}(\mathbf{x}|a1))$$
$$\mathbf{diff_f}(\mathbf{f}|\mathbf{r1}) - \mathbf{diff_f}(\mathbf{f}|\mathbf{r2}) = \mathbf{diff_f}(\mathbf{f}(\mathbf{x}|a1+da)) + \mathbf{diff_f}(\mathbf{f}(\mathbf{r}|a1))$$
$$\mathbf{diff_b}(\mathbf{f}|\mathbf{r1}) - \mathbf{diff_b}(\mathbf{f}|\mathbf{r0}) = \mathbf{diff_b}(\mathbf{f}(\mathbf{x}|a1-da)) + \mathbf{diff_b}(\mathbf{f}(\mathbf{x}|a1))$$

With

$$\mathbf{diff_f}(\mathbf{f}|\mathbf{x1}) = \mathbf{diff_f}(\mathbf{f}|\mathbf{r1}) + \mathbf{diff_f}(\mathbf{f}(\mathbf{r}|a1+da))$$

one has related material, local and indexed differences.

Again no restrictions of continuity are placed on the function **f**; consequently the bridge theorem holds for any **f**, scalar or vector, whether or not continuous. However, if at **r1**(**x1**,a1) **f** suffers a discontinuity,

$$\mathbf{f}(\mathbf{r1}(\mathbf{x1},a1)) \equiv \mathbf{f01}$$

may not equal

$$\mathbf{f11} \equiv \lim \mathbf{f}(\mathbf{r1}(\mathbf{x1},a1+da)) \text{ as } da \to 0.$$

The difference $\mathbf{diff_f}(\mathbf{f}|\mathbf{x1}) = \mathbf{f}(\mathbf{r}(\mathbf{x1},a1+da)) - \mathbf{f01})$ would then differ from **f**(**r**(**x1**,a1+da)) – **f11**). The second form holds for **f11**, not **f01**.

The theorem holds even for stalled processes.

The differences of the bridge theorem are called the **materially referenced diff(f|x)**, the **locally referenced diff(f|r)**, and the **net diff(f(r|a))**, differences respectively.

The latter is so named because

$$\mathbf{diff_f}(\mathbf{f}(\mathbf{r}|a1+da)) = -\mathbf{diff_f}(\mathbf{f}(\mathbf{x}|a1+da))$$
$$= \mathbf{diff_f}(\mathbf{f}|\mathbf{x1}) - \mathbf{diff_f}(\mathbf{f}|\mathbf{r1})$$

which refers to functional values for **x**(**r2**,a1+da) to **x**(**r1**,a1+da), that is the *net* change of values, **f**(**x1**(**r2**,a1+da)) – **f**(**r1**(**x2**,a1+da)), at observation a1+da. For stalled processes the net change is zero."

It should noted, however, that while

$$\mathbf{diff_f(f(r}|a1+da)) + \mathbf{diff_f(f(x}|a1+da)) = \mathbf{0}$$

$$\mathbf{diff_f(f(r}|a1)) + \mathbf{diff_f(f(x}|a1))$$
$$= \mathbf{f(r2(y2},a1)) - 2*\mathbf{f(r1(x1},a1)) + \mathbf{f(s2(x2},a1)).$$

In the backward sense,

$$\mathbf{diff_b(f(r}|a1-da)) = -\mathbf{diff_b(f(x}|a1-da))$$
$$= \mathbf{diff_b(f|x1}) - \mathbf{diff(f|r1})$$

and

$$\mathbf{diff_b(f(r}|a1)) + \mathbf{diff_b(f(x}|a1))$$
$$= -\mathbf{f(r0(y0},a1)) + 2*\mathbf{f(r1(x1},a1)) - \mathbf{f(s0(x0},a1)).$$

Thus the same material difference generates two distinct local and two distinct net differences just as the same local difference generates two distinct material and two distinct net differences.

The importance of the bridge theorem lies precisely in the relationship it expresses between an indexed change in a function (which implies more than one observation) with a fixed local change (which implies only one observation).

This perspective can be appreciated by applying the second form to **r**(**x1**,a), the changing location of the particle **x1**. Since diff(**r**(**x1**,a)|**r**) = **0**,

$$\mathbf{diff(r|x1}) = \mathbf{diff(r}|a1+da) = \mathbf{diff(r(x}|a1))$$

that is

$$\mathbf{r(x1},a1+da) - \mathbf{r(x1},a1)$$
$$= \mathbf{r2(x},a1+da) - \mathbf{r1(x},a1+da)$$
$$= \mathbf{r(x2},a1) - \mathbf{r(x1},a1)$$

where the first difference involves two observations while the latter two involve a single observation.

Now further, given an ordered set of real indexed observations with corresponding function, **f**(**r**(**x**,a)), results depending on continuity can be written.

First suppose **f** continuous at **r1**(**x1**,a1) with derivatives which perhaps may be discontinuous in either ***r***, ***x***, or *a*. Then, for example, from the second form

lim $\mathbf{diff}_f$(**f**|**x1**)/da

  = lim $\mathbf{diff}_f$(**f**|**r1**)/da + lim $\mathbf{diff}_f$(**f**(**r**|a1+da))/da, as da→0

implies

  D[a1](**f**|**x1**;d_fa) = D[a1](**f**|**r1**;d_fa) + D[**r1**,**r2**](**f**|a1+da;da)

where the derivatives are taken as basic, and

D[a1](**f**|**x1**;d_fa) ≡ lim (**f**(**r**(**x1**,a1+da)) – **f**(**r**(**x1**,a1)))/da
D[a1](**f**|**r1**;d_fa) ≡ lim (**f**(**x**(**r1**,a1+da)) – **f**(**x**(**r1**,a1)))/da
D[**r1**,**r2**](**f**|a1+da;da) ≡ lim (**f**(**r2**(**x**,a1+da)) – **r1**(**x**,a1+da)))/da.

The last derivative, which clearly differs from the first two, would be more accurately symbolized as

  lim D[**r1**,**r**(**x1**,a1+da)](**f**|a1+da;da) as da→ 0.

Since the bridge theorem holds for all differences, it likewise holds as a limit where these exist. Assuming now these limits exist (which implies more than one observation) at least in a forward or backward sense, the second form of the bridge theorem can be used to obtain derivatives.

Corollary (the first indexed corollaries to the bridge theorem.)

Given

  r1(**x1**,a1), the location
    of the reference particle **x1**(**r1**,a1);
  f(**r**(**x**,a)) continuous almost everywhere
   with forward and backward indexed derivatives
    at **r1**(**x1**,a1);

then

D[a1](**f**|**x1**;d_fa) = D[a1](**f**|**r1**;d_fa) + D[**r1**,**r2**](**f**|a1+da;da)
D[a1](**f**|**x1**;d_fa) = D[a1](**f**|**r2**;d_fa) + D[**r1**,**r2**](**f**|a1;da)
D[a1](**f**|**r1**;d_fa) = D[a1](**f**|**x1**;d_fa) + D[**x1**,**x2**](**f**|a1+da;da)
D[a1](**f**|**r1**;d_fa) = D[a1](**f**|**x2**;d_fa) + D[**x1**,**x2**](**f**|a1;da)

D[a1](**f**|**x1**;d_ba) = D[a1](**f**|**r1**;d_ba) + D[**r0**,**r1**](**f**|a1–da;da)
D[a1](**f**|**x1**;d_ba) = D[a1](**f**|**r0**;d_ba) + D[**r0**,**r1**](**f**|a1;da)

D[a1](**f**|**r1**;d_ba) = D[a1](**f**|**x1**;d_ba) + D[**x0**,**x1**](**f**|a1–da;da)
D[a1](**f**|**r1**;d_ba) = D[a1](**f**|**x0**;d_ba) + D[**x0**,**x1**](**f**|a1;da).

Thus
D[**r1**,**r2**](**f**|a1+da;da) = –D[**x1**,**x2**](**f**|a1+da;da)
D[**r0**,**r1**](**f**|a1–da;da) = –D[**x0**,**x1**](**f**|a1–da;da).

D[a1](**f**|**x1**;d_fa) + D[**x1**,**x2**](**f**|a1+da;da)
= D[a1](**f**|**x2**;d_fa) + D[**x1**,**x2**](**f**|a1;da)
D[a1](**f**|**r1**;d_fa) + D[**r1**,**r2**](**f**|a1+da;da)
= D[a1](**f**|**r2**;d_fa) + D[**r1**,**r2**](**f**|a1;da)
D[a1](**f**|**x1**;d_ba) + D[**x0**,**x1**](**f**|a1–da;da)
= D[a1](**f**|**x0**;d_ba) + D[a1][**x0**,**x1**](**f**|a1;da)
D[a1](**f**|**r1**;d_ba) + D[**r0**,**r1**](**f**|a1–da;da)
= D[a1](**f**|**r0**;d_ba) + D[**r0**,**r1**](**f**|a1;da).

The corollary applies to step functions even when **f** is a step function at **r1**(**x1**,a1).

For instance let
f(**r1**(**x1**,a1)) = **f**(**x1**(**r1**,a1) ≡ **f11**∗ua(a–a1)
f(**r2**(**x1**,a1+da)) = **f**(**x1**(**r2**,a1+da)) ≡ **f21**∗ua(a–a1)
f(**r1**(**x2**,a1+da)) = **f**(**x2**(**r1**,a1+da)) ≡ **f12**∗ua(a–a1)
f(**r2**(**y2**,a1)) = **f**(**y2**(**r2**,a1)) ≡ **f22y**∗ua(a–a1)
f(**s2**(**x2**,a1)) = **f**(**x2**(**s2**,a1)) ≡ **fs22**∗ua(a–a1)
f(**r0**(**x1**,a1–da)) = **f**(**x1**(**r0**,a1–da) ≡ **f01**∗ua(a–a1))
f(**r1**(**x0**,a1–da)) = **f**(**x0**(**r1**,a1–da) ≡ **f10**∗ua(a–a1))
f(**r0**(**y0**,a1)) = **f**(**y0**(**r0**,a1) ≡ **f00y**∗ua(a–a1))
f(**s0**(**x0**,a1)) = **f**(**x0**(**s0**,a1) ≡ **fs00**∗ua(a–a1))
where **fij** are constant vectors and ua is the unit observational step function. Then,
D[a1](**f**|**x1**;d_fa)= (**f21**–**f11**)∗D[a1](ua(a–a1)|**x1**;da)
D[a1](**f**|**r1**;d_fa) = (**f12**–**f11**)∗D[a1](ua(a–a1)|**r1**;da)
D[a1](**f**|**x2**;d_fa) = (**f12**–**fs22**)∗D[a1](ua(a–a1)|**x2**;da)
D[a1](**f**|**r2**;d_fa) = (**f21**–**f22y**)∗D[a1](ua(a–a1)|**r2**;da)
D[**r1**,**r2**](**f**|a1+da;da) = (**f21**–**f12**)∗D[a1](ua(a–a1)|a1+da;da)
D[**r1**,**r2**](**f**|a1;da) = (**f22y**–**f11**)∗D[a1](ua(a–a1)|a1;da)

D[**x1**,**x2**](**f**|a1+da;da) = (**f12–f21**)∗D[a1](ua(a–a1)|a1+da;da)
D[**x1**,**x2**](**f**|a1;da) = (**fs22–f11**)∗D[a1](ua(a–a1)|a1;da)
D[a1](**f**|**x1**;d$_b$a) = –(**f11–f01**)∗D[a1](ua(a–a1)|**x1**;da)
D[a1](**f**|**r1**;d$_b$a) = –(**f11–f10**)∗D[a1](ua(a–a1)|**r1**;da)
D[a1](**f**|**x0**;d$_b$a) = –(**fs00–f10**)∗D[a1](ua(a–a1)|**x0**;da)
D[a1](**f**|**r0**;d$_b$a) = –(**f00y–f01**)∗D[a1](ua(a–a1)|**r0**;da)
D[**r0**,**r1**](**f**|a1–da;da) = –(**f10–f01**)∗D[a1](ua(a–a1)|a1–da ;da)
D[**r0**,**r1**](**f**|a1;da) = –(**f11–f00y**)∗D[a1](ua(a–a1)|a1 ;da)
D[**x0**,**x1**](**f**|a1–da;da) = –(**f01–f10**)∗D[a1](ua(a–a1)|a1–da ;da)
D[**x0**,**x1**](**f**|a1;da) = –(**f11–fs00**)∗D[a1](ua(a–a1)|a1 ;da).

The first indexed corollaries then become transparently
(**f21–f11**)∗D[a1](ua(a–a1)|**x1**;d$_f$a)
= (**f12–f11**)∗D[a1](ua(a–a1)|**r1**;d$_f$a)
+ (**f21–f12**)∗D[a1](ua(a–a1)|a1+da;d$_f$a)
(**f21–f11**)∗D[a1](ua(a–a1)|**x1**;d$_f$a)
= (**f21–f22y**)∗D[a1](ua(a–a1)|**r2**;d$_f$a)
+ (**f22y–f11**)∗D[a1](ua(a–a1)|a1;d$_f$a)
(**f12–f11**)∗D[a1](ua(a–a1)|**r1**;d$_f$a)
= (**f21–f11**)∗D[a1](ua(a–a1)|**x1**;d$_f$a)
+ (**f12–f21**)∗D[a1](ua(a–a1)|a1+da;d$_f$a)
(**f12–f11**)∗D[a1](ua(a–a1)|**r1**;d$_f$a)
= (**f12–fs22**)∗D[a1](ua(a–a1)|**x2**;d$_f$a)
+ (**fs22–f11**)∗D[a1](ua(a–a1)|a1;d$_f$a)
–(**f11–f01**)∗D[a1](va(a–a1)|**x1**;d$_b$a)
= –(**f11–f10**)∗D[a1](va(a–a1)|**r1**;d$_b$a)
– (**f10–f01**)∗D[a1](va(a–a1)|a1–da;d$_b$a)
–(**f11–f01**)∗D[a1](va(a–a1)|**x1**;d$_b$a)
= –(**f00y–f01**)∗D[a1](va(a–a1)|**r0**;d$_b$a)
– (**f11–f00y**)∗D[a1](va(a–a1)|a1;d$_b$a)
–(**f11–f10**)∗D[a1](va(a–a1)|**r1**;d$_b$a)
= –(**f11–f01**)∗D[a1](va(a–a1)|**x1**;d$_b$a)
– (**f01–f10**)∗D[a1](va(a–a1)|a1–da;d$_b$a)
–(**f11–f10**)∗D[a1](va(a–a1)|**r1**;d$_b$a)
= –(**fs00–f10**)∗D[a1](va(a–a1)|**x0**;d$_b$a)
– (**f11–fs00**)∗D[a1](va(a–a1)|a1;d$_b$a).

Corollary (the third indexed corollaries to the bridge theorem)

Given

r1(**x1**,a1), the location of the reference particle **x1**(**r1**,a1);

f(**r**(**x**,a)) continuous almost everywhere with continuous gradient in section **Urn** at **r1**(**x1**,a1) into which D[a1](**r**|**x1**;d_na) extends;

then

D[a1](**f**|**x1**;d_fa)
= D[a1](**f**|**r1**;d_fa)
+ D[a1](**r**|**x1**;d_fa)•**T[URf**(a1+da)$^{-1}$**]•[URf**(a1+da)**]**$^{-1}$
•**T[D3[r1]∗(f**|a1+da;**dRf**)**]**

D[a1](**f**|**x1**;d_fa)
= D[a1](**f**|**r2**;d_fa)
+ D[a1](**r**|**x1**;d_fa)•**T[URf**(a1)$^{-1}$**]•[URf**(a1)**]**$^{-1}$
•**T[D3[r1]∗(f**|a1;**dRf**)**]**

D[a1](**f**|**r1**;d_fa)
= D[a1](**f**|**x1**;d_fa)
+ D[a1](**x**|**r1**;d_fa)•**T[UXf**(a1+da)$^{-1}$**]•[UXf**(a1+da)**]**$^{-1}$
•**T[D3[x1]∗(f**|a1+da;**dXf**)**]**

D[a1](**f**|**r1**;d_fa)
= D[a1](**f**|**x2**;d_fa)
+ D[a1](**x**|**r1**;d_fa)•**T[UXf**(a1)$^{-1}$**]•[UXf**(a1)**]**$^{-1}$
•**T[D3[x1]∗(f**|a1;**dXf**)**]**

D[a1](**f**|**x1**;d_ba)
= D[a1](**f**|**r1**;d_ba)
+ D[a1](**r**|**x1**;d_ba)•**T[URb**(a1–da)$^{-1}$**]•[URb**(a1–da)**]**$^{-1}$
•**T[D3[r1]∗(f**|a1–da;**dRb**)**]**

D[a1](**f**|**x1**;d_ba)
= D[a1](**f**|**r0**;d_ba)
+ D[a1](**r**|**x1**;d_ba)•**T[URb**(a1)$^{-1}$**]•[URb**(a1)**]**$^{-1}$
•**T[D3[r1]∗(f**|a1;**dRb**)**]**

D[a1](**f**|**r1**;d_ba)
= D[a1](**f**|**x1**;d_ba)
+ D[a1](**x**|**r1**;d_ba)•**T[UXb**(a1–da)$^{-1}$**]•[UXb**(a1–da)**]**$^{-1}$
•**T[D3[x1]∗(f**|a1–da;**dXb**)**]**

D[a1](**f**|**r1**;d_ba)
= D[a1](**f**|**x0**;d_ba)
+ D[a1](**x**|**r1**;d_ba)•**T[UXb**(a1)$^{-1}$]•[**UXb**(a1)]$^{-1}$
•**T[D3[x1]**∗(**f**|a1;**dXb**)]

which imply

D[a1](**r**|**x1**;d_fa)•**T[URf**(a1+da)$^{-1}$]•[**URf**(a1+da)]$^{-1}$
•**T[D3[r1]**∗(**f**|a1+da;**dRf**)]
= –D[a1](**x**|**r1**;d_fa)•**T[UXf**(a1+da)$^{-1}$]•[**UXf**(a1+da)]$^{-1}$
•**T[D3[x1]**∗(**f**|a1+da;**dXf**)]

D[a1](**r**|**x1**;d_ba)•**T[URb**(a1–da)$^{-1}$]•[**URb**(a1–da)]$^{-1}$
•**T[D3[r1]**∗(**f**|a1–da;**dRb**)]
= –D[a1](**x**|**r1**;d_ba)•**T[UXb**(a1–da)$^{-1}$]•[**UXb**(a1–da)]$^{-1}$
•**T[D3[x1]**∗(**f**|a1–da;**dXb**)].

The above results reduce to the continuous cases as described below.

The Continuous Case of the Bridge Theorem

The corollaries to the bridge theorem hold even when **f** and its derivatives enjoy only a limited continuity at **r1**(**x1**,a1). Where **f** and its derivatives are simply continuous simpler forms may be written. In the simply continuous case forward and backward derivatives are equal.

Corollary (the first indexed corollaries to the bridge theorem, simply continuous case)

D[a1](**f**|**x1**;da) = D[a1](**f**|**r1**;da) + D[a1](**f**(**r**|a1+da);da)
D[a1](**f**|**x1**;da) = D[a1](**f**|**r1**;da) – D[a1](**f**(**x**|a1+da);da)
D[a1](**f**|**x1**;da) = D[a1](**f**|**r2**;da) + D[a1](**f**(**r**|a1);da)
D[a1](**f**|**x2**;da) = D[a1](**f**|**r1**;da) – D[a1](**f**(**x**|a1);da)

implying

D[a1](**f**(**r**|a1+da);da) = –D[a1](**f**(**x**|a1+da);da)

D[a1](**f**|**x1**;da) + D[a1](**f**(**x**|a1+da);da)
= D[a1](**f**|**x2**;da) + D[a1](**f**(**x**|a1);da)
D[a1](**f**|**r1**;da) + D[a1](**f**(**r**|a1+da);da)
= D[a1](**f**|**r2**;da) + D[a1](**f**(**r**|a1);da).

Corollary (the third indexed corollaries to the bridge theorem, simply continuous case)
D[a1](**f**|**x1**;da)
= D[a1](**f**|**r1**;da)
+ D[a1](**r**|**x1**;da)•**T[D3[r1]**(**f**|a1+da;**dr**)]
D[a1](**f**|**x1**;da)
= D[a1](**f**|**r1**;da)
– D[a1](**x**|**r1**;da)•**T[D3[x1]**(**f**|a1+da;**dx**)]
D[a1](**f**|**x1**;da)
= D[a1](**f**|**r2**;da)
+ D[a1](**r**|**x1**;da)•**T[D3[r1]**(**f**|a1;**dr**)]
D[a1](**f**|**x2**;da)
= D[a1](**f**|**r1**;da)
– D[a1](**x**|**r1**;da)•**T[D3[x1]**(**f**|a1;**dx**)]
which imply
D[a1](**r**|**x1**;da)•**T[D3[r1]**(**f**|a1+da;**dr**)]
= –D[a1](**x**|**r1**;da)•**T[D3[x1]**(**f**|a1+da;**dx**)].

www.ingramcontent.com/pod-product-compliance
Lightning Source LLC
Chambersburg PA
CBHW071435210326
41597CB00020B/3804
* 9 7 8 1 6 3 6 2 5 6 3 9 9 *